Computational Auditory Scene Analysis

Computational Auditory Scene Analysis

Edited by

David F. Rosenthal
Lernout and Hauspie Speech Products, Inc.
Burlington, Massachusetts

Hiroshi G. Okuno
Nippon Telegraph and Telephone Corporation
Japan

CRC Press
Taylor & Francis Group
Boca Raton London New York

CRC Press is an imprint of the
Taylor & Francis Group, an **informa** business

First published 1998 by Lawrence Erlbaum Associates, Inc.

Published 2019 by CRC Press
Taylor & Francis Group
6000 Broken Sound Parkway NW, Suite 300
Boca Raton, FL 33487-2742

© 1998 by Taylor & Francis Group, LLC
CRC Press is an imprint of Taylor & Francis Group, an Informa business

First issued in paperback 2019

No claim to original U.S. Government works

ISBN 13: 978-0-367-44784-7 (pbk)
ISBN 13: 978-0-8058-2283-0 (hbk)

**Visit the Taylor & Francis Web site at
http://www.taylorandfrancis.com**

**and the CRC Press Web site at
http://www.crcpress.com**

Library of Congress Cataloging-in-Publication Data

Computational auditory scene analysis / edited by David F. Rosenthal,
  Hiroshi G. Okuno.
    p.   cm.
    ISBN 0-8058-2283-6 (cloth : alk. paper)
    1. Automatic speech recognition. 2. Computer sound processing.
  3. Signal processing. I. Rosenthal, David F. II. Okuno, Hiroshi G.
  TK5936.S65C66 1997
  006.4'54--dc21   97-5936
                                                                  CIP

Publisher's Note
The publisher has gone to great lengths to ensure the quality of this reprint but points out that some imperfections in the original may be apparent.

Contents

Preface

The papers selected for inclusion in this collection are representative of a growing body of work in computational auditory scene analysis (CASA). Until recently, most of the work in computer understanding of sound has been heavily concentrated on the problem of automatic speech recognition (ASR), and today ASR is still a primary goal. Increasingly, however, researchers have been struck by the gap between computer ASR systems (as they are usually implemented) and the capabilities of the human auditory system. This interest is motivated not only by the desire to understand how the human auditory system works, but also by the growing awareness that unless they embody such an understanding, ASR systems will not be effective in unconstrained environments.

A fundamental characteristic of human hearing is the ability selectively to attend to sound originating from a particular physical source, in an environment that includes a variety of sources ("the cocktail party effect") (Cherry, 1953). In contrast, most computer sound-understanding systems are structured to regard the world as consisting of "signal" (usually, speech) and "noise" (everything else).

To operate effectively as listeners in the real world, computers will have to acquire a more flexible and sophisticated view of what constitutes signal, and what constitutes interference. Such observations have led the researchers whose work is represented here to try to integrate speech and non-speech recognition and understanding systems into a common framework. The framework which underlies most of these recent efforts is that outlined in *Auditory Scene Analysis*, by Albert Bregman of McGill University (Bregman, 1990). Most of the papers in this collection can be viewed as descriptions of computational implementations of the views outlined in *Auditory Scene Analysis*.

1 A BRIEF HISTORY OF CASA

The history of CASA might properly be said to have begun with some important work not in hearing, but rather in vision — namely that of David Marr (Marr, 1982). Marr presented what he called a "computational theory of vision," or what we would now call a *functional specification for the vision problem*. This is a formal statement of the problem that vision is attempting to solve, as distinguished from a particular way of solving it. The importance of distinguishing between a functional specification and an implementation has since become a fundamental tenet of software engineering; Marr's essential contribution was to give the first coherent account of a problem of computational perception in these terms.

Marr viewed vision as a sequence of representational stages, called *sketches*, beginning with representations which are structurally close to the original light pattern entering the retina, and progressing to higher levels of abstraction. The relevance of this idea to auditory and speech processing was appreciated by a number of researchers. In general, however, research in hearing was concentrated almost entirely on problems associated with speech processing; until recently there was no comprehensive theory of hearing, which would include speech processing as one of its parts (Cooke, 1991; Ellis, 1992).

It is interesting that the *"Vision"* of auditory research came not from the computational community, but rather from an experimental psychologist, Albert Bregman. His *Auditory Scene Analysis* presents the most convincing and widely accepted *functional specification of the hearing problem.* Bregman posits that the first major stage in the processing of auditory information -- which might correspond, roughly, to Marr's 2 1/2 D sketch -- is the separation of sound into *streams,* or components which correspond to individual physical sources.

In the years following the publication of *Auditory Scene Analysis* a number of researchers began to try to build "Computational ASA" systems -- that is, systems that could separate a continuous sound signal into components corresponding to its separate physical sources. That work is the core of what is represented in this collection; all of this work relates to the problem of separating sound into sources.

2 RELATED DEVELOPMENTS

Another line of inquiry relevant to the development the field is research into *blackboard architectures.* Blackboard architectures first appeared on the 1970's, in the context of an ARPA-sponsored competition to for building speech understanding systems. HEARSAY II, developed by Victor Lesser and his colleagues is often cited as the first such architecture (Erman, 1980).

Blackboard architectures are motivated by the observation that in a perceptual task such as hearing and understanding speech, there is information available at a variety of levels of abstraction -- so, for example, when we understand a spoken sentence, we are likely to be using processes which understand low-level signal-processing information to decide which formants are present, mid-level constraints on what sounds can follow others in the English language to decide what words are being spoken, and high-level knowledge of what a word is likely to have been spoken, based on the context of the conversation. All of these might bear on the question of which vowel was uttered at a given time. Since there may be error or ambiguity in any of these processes, applying them in a fixed order is likely to fail. If on the other hand, we can opportunistically use whatever level is currently providing the most reliable information, and use that level to establish a context and constrain the others, we have a much better chance of success.

In the context of the ARPA speech initiative, blackboard systems were less successful than other systems with much simpler architectures, particularly those based on statistical models of speech, and in the years following the ARPA project, most speech research came to use a statistical technique called Hidden Markov Models (HMM's). Today most commercial ASR systems are based on HMM's. A minority of researchers, however, concluded that the apparent failure of blackboard models was not due to their inherent inferiority, but rather to the fact that implementing systems with complex control structures is difficult and not yet well-understood. Furthermore, there are many perceptual tasks where statistical models are not available, so the problems of implementing more flexible architectures will eventually have to be confronted. Much of the recent blackboard-related work in hearing has been carried out by Hamid Nawab, Victor Lesser, and their students and colleagues at the Boston University and the University of Massachusetts at Amherst.

A related line of development is that of *agent-based* systems. Agent-based systems are inspired by the observation that complex, apparently intelligent behavior can sometimes *emerge* from the collective actions of relatively simple, independent modules called *agents,* rather than being explicitly architected by the system designer (Minsky, 1986). Much of the CASA-related, agent-based research has been pursued by a group led by Hiroshi Okuno of NTT Basic Research Laboratories, and associated researchers at Waseda University.

Recently, approaches which are not accurately characterized as either agent-based blackboard-based have been pursued; like blackboards and agent-based systems, these approaches also address the issue of integrating evidence from different levels of abstraction. An example is the *model-based* approach exemplified by the work of Kunio Kashino, now with NTT Basic Research Laboratories. Kashino's system for transcribing music carefully considers the confidence that should be placed on hypothetical explanations of observed acoustic data by using Bayseian belief networks.

The confluence of ideas about blackboards, agent- and model-based architectures, and research into the computational basis of hearing has been motivated by the interests of researchers from a variety of backgrounds. Researchers attempting to build computational models of hearing have concluded that systems which process information at strictly ordered levels of abstraction lack the flexibility to deal with real-world sound. Researchers who are primarily interested architectural considerations, for their part, have seen CASA as a fertile proving ground for their ideas. At the same time, the increasing desirability of structured electronic archives, more accurate ASR, and more sophisticated hearing aids have provided motivation and opportunities from the applications viewpoint.

This commonality of interest has been the basis for a number of recent international gatherings, among them the Twelfth International Pattern Recognition Conference (Jerusalem, October 1994), an Institute of Acoustics workshop on Speech Technology and Hearing Science (Sheffield, January 1995), The Computational Auditory Scene Analysis workshop at the

International Joint Conference on Artificial Intelligence (IJCAI, Montreal, August 1995) and a session at the IEEE Workshop on Applications of Signal Processing to Audio and Acoustics (Mohonk, New York, October 1995). It was at the 1995 IJCAI-CASA workshop that the papers included here were first presented.

The 24 papers in this collection are grouped into four sections: "Physiological and Neural Models," "Architecture and Control," "Representation and Signal Processing," and "Speech and Other Applications." These division correspond, very roughly, to the background and orientation of the authors. These divisions should not be interpreted as "hermetic," however; in many cases, an idea which we consider fundamental arises in several different sections. In other words, were this collection a richly-hyperlinked document (as, no doubt, it should be!) there would be almost as many section-crossing hyperlinks as within-section hyperlinks. Many of the papers might easily have fit into a different section from that in which we placed them. It is the commonality of the ideas in these papers, rather than the differences in backgrounds of their authors, which we feel is interesting and deserving of emphasis.

The IJCAI-CASA workshop and this resulting collection of papers were in every sense a group effort by the CASA community, and we'd like to thank all of the contributors. Dan Ellis provided much-needed assistance at every level, from general conception and organization to finicky software platform and version problems, and everything in between. Kim Oey-Rosenthal spent endless hours typesetting the manuscript and resolving annoying difficulties. Other people deserving of particular mention are: Martin Cooke, Hamid Nawab, Malcolm Crawford, Malcolm Slaney, Ray O'Connell, Masataka Goto, Kunio Kashino, Alon Fishbach, and Al and Abigail Bregman.

David Rosenthal
Hiroshi G. Okuno
Tokyo and Arlington, Massachusetts
1996-1997

REFERENCES

Bregman, A. S. (1990). *Auditory Scene Analysis: The Perceptual Organization of Sound.* Cambridge, MA.: The MIT Press.

Cherry, E. C. (1953) Some experiments on the recognition of speech, with one and with two ears. *Journal of Acoustic Society of America,* .25, 975-979.

Cooke, M. P. (1993). *Modelling Auditory Processing and Organisation.* Cambridge, U. K.: Cambridge University Press.

Ellis, D. P. W. (1992). A perceptual representation of audio. MS thesis, Department of Electrical Engineering and Computer Science, MIT.

Erman, L. D. , Hayes-Roth, F., Lesser, V. R., & Reddy, D. R. (1980). The Hearsay-II speech-understanding system: Integrating knowledge to resolve uncertainty. *Computing Surveys,* ACM, (12), pp.213-253.

Marr, D. (1982). *Vision.* New York: W. H. Freeman.

Minsky, M. (1986). *The Society of Mind.* New York Simon & Schuster.

1

Psychological Data and Computational ASA

Albert S. Bregman
McGill University

The strategies used in human auditory scene analysis (ASA) have implications for the architecture of a computational ASA system. To model the stability of the human ASA, the computational system must allow different cues to collaborate and compete, and must account for the propagation of constraints across the frequency-by-time field. A scaleable architecture must be designed in which the number of cues can easily be extended and the difficulty of the problems increased without requiring a change in the basic design. Unless modelers are familiar with a wide range of psychological data, they may develop architectures that are successful only on "toy" problems, or that model only a few phenomena, and do not lend themselves to being expanded to deal with a wider class. Finally, if speech recognition systems are to exploit the pre-processing done by ASA, they must be modified so they can use the data linkages that have been proposed by ASA.

1.1 INTRODUCTION

Many kinds of events in the environment give rise to acoustic pressure waves. Typically, the sounds are overlapped in time and in frequency. Yet, despite the multiplicity of concurrent sounds, there is only one acoustic pattern at the ear of the listener - the sum of the individual ones. Despite this mixing, our auditory systems can recover separate descriptions for each separate sound-producing event. I have referred to this accomplishment (Bregman, 1990) as auditory scene analysis (ASA).

Although psychology and the construction of computational ASA systems often have different goals and certainly have different methods, I believe it is important for psychologists like myself to share ideas with artificial intelligence

1

(AI) researchers in computational ASA. Computational modelers of ASA vary in their concern with how humans carry out ASA. At one extreme are system designers who attack the ASA problem with no concern for data from human ASA. At the other, there are researchers who explicitly try to model human performance with the intention of comparing the fine details in the output of their models with the data from experiments on humans.

Probably the majority of people actually working on computational ASA fall somewhere in between. Although not explicitly trying to model the human auditory system, they recognize the obvious fact that the human perceptual system actually achieves ASA in a robust way. This inclines them to try to follow the human approach to the extent that it is known (e.g., Brown, 1992).

The benefits of an alliance between AI and psychology can flow in both directions. We psychologists try to provide an account of how human ASA uses various cues and heuristics and combines evidence; we also try to show how this affects perception. Our research can lay constraints on the range of acceptable computational models.

Computational modelers can help by exposing unspoken and unrealistic assumptions embedded in the theories of psychologists. For example, years ago, when I was thinking about the grouping of spectral components undergoing parallel or nonparallel changes in frequency, I failed to consider the problems raised by the trade-off between the resolutions in frequency and time. Issues concerning resolution are very important, and often become apparent only when one is actually trying to construct a model (e.g., Mellinger, 1991).

1.2   HEURISTICS OF BOTTOM-UP (PRIMITIVE) ASA

As a psychologist, then, I review some of the principles that govern the human's organization of sound in the laboratory. There are at least two aspects to the organization imposed by ASA - simultaneous and sequential organization - which correspond to the two dimensions of a spectrogram. Simultaneous organization determines which parts of the spectrum, present at the same time, will be interpreted as parts of the same sound, whereas sequential organization determines which components, distributed over time, will become parts of the same stream.

Bottom-up ASA is accomplished by exploiting any regularity in the changing spectrum that suggests which groups of components should be treated as separate sounds. For example, if, at some moment, sets of harmonics related to two different fundamentals are detected, ASA separates them into two groups. This strategy is justified by a regularity in the environment, namely that sound is often derived from an oscillating source that generates a harmonic signal.

One can think of the auditory system as possessing a number of heuristic methods for scene analysis, each exploiting a different regularity in the environment. Here I review a number of the heuristics and regularities related to simultaneous organization.

One of these is a class of regularities that the Gestalt psychologists called common fate, which is present whenever spectral components undergo parallel changes over time, either in frequency or in amplitude or even, perhaps, in spatial location or phase. For example, parallel changes in the frequencies of components will occur when a voice glides from one pitch to another. Synchronized changes in the amplitudes of components will occur when a stop consonant simultaneously interrupts all the components (see Bregman, 1990, Chapter 3, for an overview of common fate in ASA).

These regularities in the signal are due to characteristic ways in which sound is generated in the environment. Truly unrelated sounds rarely change in frequency in a synchronized way or stop and start at exactly the same time, or move through space together. These facts justify segregating components that exhibit independent changes and integrating those that undergo parallel ones.

A number of steady-state differences among simultaneous spectral components can also increase their segregation. For example, ASA tends to segregate simultaneous components more strongly when they arrive from different directions. This is justified because all frequencies that arise from a single event tend to come from the same location.

The heuristics that group components over time also use methods that reflect regularities in the environment. For example, because sounds tend not to change their characteristics abruptly, a rapid change in some property of the signal occurs when a new sound has entered the mixture, whereas a slower, continuous change often occurs when a single sound source gradually changes its way of vibrating. Accordingly, ASA will group successive sounds with similar spectra into the same perceptual stream, perhaps rejecting other interleaved sounds into a different stream because their spectra are different.

The sequential differences that cause sounds to be assigned to separate streams include the following: the positions of spectral peaks and valleys, differences in fundamental frequency, spatial location, intensity, and possibly even in phase (see Bregman, 1990, Chapter 2, for an overview of sequential integration). Sudden changes in these indicate that a new event is beginning to dominate the spectrum. The evaluation of these features requires a comparison of spectra over time, but therein lies a problem: Sequential comparison requires that the relevant spectra be segregated from other components that may occur at the same time.

1.2.1 The old-plus-new heuristic and the formation of residuals

The story about sequential and spectral organization is made more complicated by a principle of ASA called the old-plus-new heuristic, a competition between simultaneous and sequential grouping (Bregman, 1990, Chapter 3, p. 222). Because sounds tend to continue despite the onset of new ones, whenever a spectrum A turns suddenly into another one (B) possessing more components or greater energy, spectrum B should be interpreted as a mixture of two sounds: the continuing old sound, A, and a newly starting one, C, whose properties can be revealed by removing A's effects from B. This method for segregation is very powerful because it takes the history of a spectrum into account, rather than merely its current properties.

A simple example of this heuristic illustrates the decomposition of a spectrum into two parts. A rectangular band of noise, 400 ms in duration, has frequencies from 0 to 1 kHz (the "low" burst). It is repeatedly alternated with a shorter burst, 200 ms in duration, whose spectrum extends from 0 to 2 kHz (the "wide-band" burst), with no silences between them. Listeners interpret the cycle as a steady low noise, with no breaks in it, joined every 600 ms by a short high noise burst. This is similar to a stimulus studied by Warren (1984).

The percept of the low continuous burst is derived from 0-1 kHz burst and from the 0-1 kHz frequencies extracted perceptually from the wide-band burst. The perceived high burst is the residual formed when the lower half of the wide band is taken away and given to the low stream. The old-plus-new heuristic, faced with a spectrum that suddenly contains more components, tries to interpret the wider spectrum as a mixture of the continuing narrower one with something else. Therefore it groups the low spectrum with the lower half of the wider spectrum and treats the two as a single, continuing sound. At the same time, it attempts to remove the low spectrum from the wide one, and derives, as a residual, a higher, 1-2 kHz sound, hears it as a high band of noise.

Another example of this heuristic is obtained by taking a spoken sentence, deleting every other 250 ms segment and replacing it with intense white noise so that 250 ms fragments of the sentence alternate with 250 ms bursts of noise. We experience a continuous sentence, which seems to be accompanied by a pulsing noise (see Miller and Licklider, 1950). This interpretation results from the old-plus-new heuristic: At the onset of each noise burst, new spectral components are being added by the noise. These are interpreted as the beginning of an additional sound, initiating a voice-noise mixture. The noise components that match the expected part of the sentence are allocated to the voice stream.

This is an interesting example of the joint functioning of bottom-up and top-down organization. The bottom-up effects include the requirement for continuity, that is that there be no sudden change in the A sound just before and after the B sound to suggest that the simpler spectrum is going on and off. The top-down process contributes the sounds of particular words that are heard as continuing under the noise.

1.3 ARCHITECTURE OF AN ASA SYSTEM

It appears to me that there are easier and harder parts in the computational modeling of ASA. Although it might not seem so to some who have tried it, I think the easiest part is building mechanisms to extract the cues from the incoming data. It is a good deal harder to figure out how to make them interact properly to converge on a good grouping of the sense data, because this involves choosing an architecture for the system. I propose to lay out a set of requirements that, from my perspective as a psychologist, should govern this choice of architecture.

Units and Features. The first is the issue of units. When we say that two kinds of cues collaborate in the control of grouping, we should ask, "the grouping of what?". Does the ungrouped data consist of primitive units? Is it even necessary to think in terms of units? Because we use the word, grouping, in describing the activity of ASA, we tend to think in terms of a first-level analysis that breaks down the signal into elements that can later be grouped. But it is possible that the elements, or units, come out of the analysis, rather than being given beforehand.

As if this were not complex enough, it may be that the computation of features may take place more than once, at different levels. It may be done once on the raw continuum, before any units are extracted, in order to decide on the nature of the units, and then may take place again on the units, to see whether they resemble other units. To add to the complexity, it may be that these analyses go on concurrently.

Redundancy and Stability. If a model is to work as human ASA does, increasing the number of available cues should make it perform better, and taking away any particular type of cue should not trip it up, but should merely reduce its performance, since the model must behave in a stable manner across a range of situations where particular cues may or may not be available.

It is fortunate that many cues and heuristics of ASA have been identified, because none of them can be trusted all the way, due to the fact that any particular cue may be blurred or absent in a particular environment. For instance, the harmonicity heuristic will be ineffective for sounds that do not have a harmonic structure. Heuristics based on spatial location may be ineffective when the environment is very reverberant or when the sound is blocked from reaching one of the listener's ears. Because no cue is always available, different ones must be allowed to compete and cooperate in the organization of the signal. Therefore, the architecture of the system should allow evidence provided by different cues to be added up easily in determining the final organization.

Competition. It is also important that the model exhibit the competition that occurs between forces of sequential and simultaneous integration. For example harmonicity cues may suggest that an acoustic component belongs to a simultaneous group of partials because they are all related to the same fundamental. At the same time the fact that this component is a smooth continuation of an earlier isolated pure tone may activate a sequential grouping that conflicts with the simultaneous one.

Competition and collaboration does not only occur between different acoustic properties of in the signal, but also when the same relation occurs at different points in the frequency-by-time field. For example, in frequency-based grouping, two components, X and Y, may group sequentially only because their proximity in frequency to one another is greater than to other tones. However, if a third component, Z, is added, very close to Y in frequency, Y may no longer group with X, preferring, instead, to group with the nearer tone, Z. Here we see a competition among the instances of a single relation, frequency proximity, at different places in the frequency-by-time field.

So relative proximity, not just raw proximity, plays a role in grouping. To deal with this fact, the model should clarify, in a general way, how competition works in establishing an organization. The early analysis of signals would have to yield "fields" of supportive and competitive relations among components of the signal. Some relaxation methods in computer vision use fields of this sort.

Propagation of Constraints. Apparently, in ASA, constraints can propagate in this field (Bregman and Tougas, 1989). For example, suppose we alternate a pure tone, A, with a mixture of two pure-tones, B and C. If A is close enough in frequency to B, it can strip it out of the BC mixture into a repeating AB sequence. However, it is not merely the relation between A and B that governs the AB grouping. One can insert a tone after the BC sequence, near in frequency to C. The repeating sequence is now A B/C D, A B/C D.... If D is close enough in frequency to C it captures it into a repeating CD stream, releasing B to group with A. So the effects of the BC grouping have propagated across the field to affect the AB grouping. Relaxation labeling methods also show this behavior.

1.3.1 Designing a scaleable architecture

I now discuss a different sort of constraint on the architecture of an ASA system: It should scale up with no basic change in structure. Therefore it should be designed from the beginning with an appreciation of what is involved as things get more complex.

There are two aspects to this scalability. The first is that one should be able to incorporate cues other than the ones with which the system was first tested, without changing its architecture.

The second is that the methods should still work when the problems are scaled up from "toy" problems, in which the constraints on the incoming data are severe, to more complex problems with less constrained data. For example, some ASA systems have been tested on a pair of sentences composed of vocalic sounds only, such as "We were away when you were." This constraint makes it possible to rely primarily on the detection of harmonicity. However, harmonicity in environmental sounds is restricted primarily to animal vocalizations and the flight of insects. Although such signals are very important to the human listener, so are other sounds.

It is perfectly acceptable to limit a problem to one that can be solved using only a few heuristics, but researchers must be critical and ask whether the system's architecture allows the addition of more heuristics in a natural way. If the present architecture were to remain, would the system ever be able to deal with mixtures of whispering voices, or more than two voices, or with mixtures of sounds that are unlike the human voice, such as footsteps and the rustling of leaves in the wind? More importantly, will it work when the type and number of sounds is not given in advance?

The choice of an architecture, rather than the cues to be used, should be given top priority in the planning process. Researchers investing a great deal of effort in refining the capacities of a particular architecture using a few heuristics on a toy environment may discover that they have been going down a blind alley and must totally redesign the architecture if they wish to deal with a less constrained acoustic environment.

We have so far concerned ourselves with models that attempt to solve the ASA problem directly. There is, however, another approach: trying to model the data that comes out of the perception laboratory. This is a dangerous mission and again requires a wide knowledge of ASA phenomena. Without it, a researcher may invest a lot of effort to develop a model that offers a parsimonious account of a very limited subset of laboratory phenomena. Consequently, while the model may be very parsimonious in accounting for a few perceptual effects, it may turn out to be so specific to that small set of phenomena that it is helpless when a wider range of laboratory effects has to be explained. Again, an early stage in the development of a model of this type should be to ask whether it is too narrowly focused.

Models versus Theories. A formal model is often thought of as an algorithm for generating the results of experiments. However, when the range of experiments is small and the number of computer instructions large, the ability to mimic data is not a convincing proof of the adequacy of the underlying theory. Indeed, in some cases, there is no theory at all underlying the model, but merely a set of computations that reproduce the shape of the data. Such models can be seen merely as capturing some of the formal properties of the data, rather than explaining the phenomena. They are simply data-reduction techniques. A model is not automatically a explanatory theory; it is theories that are of ultimate importance, not models, even for practical purposes.

1.3.2   Relation between top-down and bottom-up constraints

Another issue that affects the choice of an architecture is the relation between bottom-up and top-down constraints in ASA. The function of ASA is to inhibit the building of a description of a "chimerical" event, derived by combining properties drawn from a number of different environmental events. But it is not necessary for ASA to partition the sense data in order to do so.

Models that Derive Two Separate Signals. The goal of partitioning data preliminary to recognition has led some researchers to test their bottom-up models by getting them to take the incoming data apart into separate streams, and then to acoustically reconstruct separate signals. These are then played out separately to the researcher who, it is assumed, will someday be replaced by a computational model of recognition.

Although this is useful for getting a crude idea of how well the computer's ASA processes are doing, these models isolate the component signals in a much more rigid way that the human ASA process does. There seems to be no purely bottom-up process in humans that packages the data into airtight compartments representing individual sound sources. This conclusion is forced on us by two phenomena, the effects of "trying" to hear the sounds in a particular way, and certain observations in speech perception.

Trying. The role of "trying" may be observed at certain speeds at which a sequence of sounds may be heard either as a single stream or as two streams, depending on the intentions of the listener. If it were strictly true that the segregations made by bottom-up ASA could not be overcome by top-down processes, then trying could only affect stream segregation if the process of trying were a bottom-up process. Yet trying must be top-down, because it is controlled consciously.

Duplex Perception of Speech. The phenomenon of duplex perception of speech (DPS) also challenges the idea that bottom-up organization creates airtight packages that can only be examined one at a time by pattern-recognition processes. Suppose two syllables, simplified versions of /da/ and /ga/, are distinguished only by a short third-formant transition (Mattingly et al., 1971). Suppose, further, that this distinguishing transition is played to one ear over headphones when the rest of the syllable (called the base) is played to the other ear, with transition and base temporally aligned as in the original syllable. Sometimes the transition presented with the base is that of /da/, and sometimes /ga/.

Listeners report a strange experience. They hear a complete syllable coming from the ear that is receiving the base. But its identity (/da/ versus /ga/) depends on which transition is being presented to the other ear. This shows that phonetic recognition, presumably a top-down process, can combine the information from the two ears. Yet the listeners also hear the transition as a separate sound located at the ear to which it is presented. This implies that bottom-up processes, based on spatial perception, have segregated the transition from the base. Here we have a case in which bottom-up ASA seems to have created two packages of information whereas a top-down recognition process has taken data from both packages, apparently defeating the original purpose of bottom-up ASA.

Implications for Computational Models of ASA. DPS suggests that, at least in the perception of speech, we cannot view segregation as a complete packaging preliminary to recognition. There are two ways to deal with this difficulty. The simpler is to assume that although bottom-up ASA acts first, the segregation it imposes is not all-or-none, but can bind components with links that vary in strength. If the links are not too tight, recognition processes, looking for specific sets of components, can select components despite their links to other ones.

Another option is to consider bottom-up and top-down processes as operating together, so that the description-forming process tries to satisfy both bottom-up and top-down constraints at the same time. Either option implies a weaker role for bottom-up grouping than my earlier theorizing and some computational models have depicted, and calls into question the strategy employed by the some models of using bottom-up cues to achieve such an absolute partitioning of the auditory data that the original signals can be reconstructed. The goal of such models is to use ASA as a front end to recognition processes, which would then operate on the results of the bottom-up analysis. This approach would make it unnecessary for the recognition models to be redesigned to accept linked data as input rather than signals, but I doubt whether human ASA works this way.

Is Speech Perception Immune to Primitive ASA? A third approach to the data from DPS is to argue that only non-speech recognition processes accept the all-or-none partitioning of sense data, but speech perception is exempt. This option must be examined by both perceptual psychologists and modelers.

Another observation that favors this idea is that ordinary listening to speech seems to contradict the idea that sequential similarity affects the grouping of components. For example, in a repeating sequence of sounds, if you rapidly alternate a tone with a burst of noise, the noise will segregate from the tone. Yet in natural speech, a noisy segment may be preceded and followed by a periodic segment as in the word nation. The two types of sound must be integrated if the word is to be recognized. How can this difference be reconciled?

The speech-is-special theorists can argue that speech perception has its own rules for the integration of sounds, perhaps due to its access to knowledge about how speech is articulated. Another possible interpretation would argue that a in normal speech perception, the strong sequential expectation, built up through experience with the language, overcome any segregative tendencies. Either of these views makes bottom-up ASA almost irrelevant in language perception.

It is hard for me to accept the irrelevance of bottom-up processes when we know that that differences in pitch and spatial location actually do make it easier to segregate two voices perceptually and track the speech of only one of them, facts that are consistent with ASA.

Repeating Cycles. Perhaps the meaning of the weak effects of perceptual differences is not the special status of speech. We must remember that laboratory studies often use repeating cycles of sounds in order to drive the organization based on a single cue to unusually high levels. Therefore in nonrepeating sequences, such as speech, we should not expect such strong effects of a single acoustic difference. Also, even when we use repeating cycles, very strong organization emerges only after a number of repetitions. In normal life situations, no cue has its strength raised to such heights by repetition. This lack of dominance by a single cue may allow the system to weigh many cues in arriving at its decisions about grouping.

Grouping is Competitive. Another thing to remember when we see differences between natural speech and cyclically repeating segments is that low-level grouping is competitive. Two sounds may group together when there is nothing better for each of them to group with. In normal speech, the next instance of the same sound may not follow soon enough for the two to be grouped together. So the grouping, in the word *nation*, of the fricative with the vocalic parts may just follow from the fact that there are no better nearby sounds for the fricative to group with.

Cumulation of Strength through Repetition. Another point is that the strong streaming, often found when alternating two categories of sound in the laboratory, depends on the cumulative build-up of strength after many repetitions. In speech, however, the sounds are drawn from more than two categories. This means that the reappearance of the same category (say /s/) will rarely appear with the very short delays that are typical of laboratory cycles, and even if it does, the event will not be repeated often enough to set up a separate stream for the /s/s.

Continuity in Speech. We should also take into account the fact that in speech, the articulators do not snap instantly from one setting to another (except, perhaps, in the case of stop consonants). It has been shown that the tendency of successive speech parts to hold together depends, in part, on the transitions created by smooth changes.

Many of these reasons for a speech sequence holding together no longer apply when two people are talking at once. For example, the likelihood increases that a sound may find a best partner in the other voice. This may account for the finding that the segregation of two concurrent voices requires (and uses) ASA principles.

Perhaps the many differences that I have listed between laboratory cycles and running speech allows us to see why groupings in natural speech are weaker. ASA may, of course, still be playing a role if we assume that its links can vary in strength. Even weak links could have a constraining effect on later recognition processes, especially with the high levels of interference that occur in mixed speech.

Are speech-specific explanations ruled out? Do these arguments imply that knowledge of articulation and the development of speech-specific expectations can be ruled out in favor of low-level ASA? Not at all. Usually, when somebody asks me which of a number of explanations are valid for some psychological observation, I answer "all of them." Like Sigmund Freud who spoke of "overdetermination" of behavior, I believe that there is great deal of redundancy in the mechanisms responsible for many phenomena in psychology, including perception.

1.4 SUMMARY

I have presented an outline of ASA that places several constraints on a computational model: (a) The model should employ the old-plus-new heuristic; (b) it should incorporate collaboration and competition in a natural way, since these two forces are at the heart of ASA; (c) it should allow for the propagation of constraints across the auditory field; (d) it should grow in ability as more heuristics of grouping are added and degrade gracefully when particular cues are removed; and (d) the outcome of ASA should be a set of data links that vary in strength.

I have even had the audacity to offer the idea that unless modelers are familiar with a wide range of psychological data, they may develop architectures that are successful only on "toy" problems, or that only model only a few phenomena, and do not lend themselves to being expanded to deal with a wider class.

This may be viewed as a challenge to computational modelers to emulate the flexibility of the human auditory system in designing their models. Of course it is easy enough for me to say this. As a psychologist, I am only expected to uncover the general approaches taken by the auditory system in solving the ASA problem. The task facing the creator of an AI model, to actually make the thing work, is much more difficult.

ACKNOWLEDGMENTS

Support is acknowledged from the Formation de Chercheurs et a l'Aide de Recherche (FCAR) program of the Province of Quebec, the Natural Sciences and Engineering Research Council of Canada, and the National Institute of Mental Health grant no. MH52254-02.

REFERENCES

Bregman, A. S. (1990). *Auditory Scene Analysis: The Perceptual Organization of Sound.* Cambridge, MA.: The MIT Press.

Bregman, A. S. & Tougas, Y. (1989). Propagation of constraints in auditory organization. *Perception and Psychophysics*, 46 (4), 395-396.

Brown, G. J. (1992). Computational auditory scene analysis: A representational approach. Ph.D. dissertation, University of Sheffield. Sheffield, England.

Mattingly, I. G., Liberman, A. M., Syrdal, A., & Halwes, T. (1971). Discrimination in speech and non-speech modes. *Cognitive Psychology*, 2, 131-157.

Mellinger, D. K. (1991). Event formation and separation in musical sound. Ph.D. Dissertation, Stanford University, Center for Computer Research in Music and Acoustics. Dept. of Music. Report no. STAN-M-77.

Miller, G. A., & Licklider, J. C. R. (1950). Intelligibility of interrupted speech. *Journal of the Acoustical Society of America*, 22, 167-173.

Warren, R. M. (1984). Perceptual restoration of obliterated sounds. *Psychological Bulletin*, 96, 371-383.

2

**A Prototype Speech
Recognizer based on
Associative Learning
and Nonlinear Speech
Analysis**

Jean Rouat and Miguel Garcia
Université du Québec à Chicoutimi

The grouping of sounds has been shown to be partially based on
amplitude modulation (AM) characteristics (Bregman et al., 1985),
suggesting that AM information observed in auditory nerve fibers
could be used by the auditory system to segregate speech from
background noise. This chapter proposes a speech recognizer
prototype that relies on patterns of modulation observed in auditory
fibers influenced by a summation of close harmonics. The
recognition task is performed by a modified Dystal (DYnamically
STable Associative Learning) neural network (Alkon et al., 1990)
(Blackwell et al., 1992). Preliminary results indicate that the
approach might be efficient and powerful. Further experiments have
still to be done in order to evaluate the approach. Experiments on
continuous speech in noisy environment are planned.

2.1 INTRODUCTION

The analysis and the recognition of speech spoken in noisy environment is a
difficult and crucial task. Most of the speech recognizers alleviate the
difficulties of this task by training on noisy data, assuming that the statistical
properties of the noise will remain unchanged between the learning and the
recognition phase. Other techniques assume that the speech source and the
interference noise sources are spatially different which is not necessarily so. In
summary, the most effective techniques are generally useful for specific
conditions.

On the other hand, perceptive analysis and auditory models enhance the
discriminant information from non stationary noise and are supposed to yield

good performance in adverse conditions. However, their complexity and the difficulty of exploiting the dynamic output information with standard pattern recognition algorithms restrict their integration in speech recognizers.

According, to Bregman (Bregman, 1984), the phasic analysis performed by the auditory system, in conjunction with the tonal analysis, is adapted to the perception of speech in adverse environment. The spectral integration (or grouping of sounds) was shown to be partially based on common amplitude modulation characteristics (Bregman et al., 1985).

Furthermore, research work on automatic demodulation of speech can be motivated by the fact that the human brain has neural cells specialized in AM and frequency modulation (FM) detection (Gardner & Wilson, 1979) (Tansley & Suffield, 1983). Robles, Ruggero and Rich (Robles et al., 1991) observed distortion products on chinchilla basilar membrane and they suggested that the living basilar membrane is a nonlinear system. Thus, the perception of distortion products could be due to the basilar membrane response and not only to neural postprocessing. Moreover, simple nonlinear operators can enhance the AM or FM information in a signal (Maragos et al., 1992) (Rouat, 1993) and can be used to process the output of a cochlea filterbank in order to obtain AM information characteristics of speech signal and segregate it from background noise (Rouat et al., 1992).

2.2 MODULATION IN THE AUDITORY SYSTEM

The modulation information is one of the main cues extracted by the auditory system. Schreiner and Langner (Langner & Schreiner, 1988) (Schreiner & Langner, 1988) showed that the inferior colliculus of the cat contains a highly systematic topographic representation of AM parameters and maps showing 'best modulation frequency' have been determined.

The poor spectral resolution of the cochlea can be an advantage when the speech signal is harmonic, as more than one harmonic of the signal can fall into the same channel producing an amplitude modulated signal with a modulation frequency equal to (or a multiple of) the fundamental frequency. Therefore, the fundamental frequency F0 (or a multiple of F0) can be encoded in the temporal discharge patterns of auditory nerve fibers. The characteristic frequencies of these nerve fibers can be very different from the fundamental frequency F0 (or a multiple of F0).

As the bandwidth of the nerve fibers vary significantly (even for fibers tuned to the same frequency) and since the bandwidth may be broad, at least some of the auditory nerve fibers might be able to encode the envelopes relevant to speech signals. Langner (Langner, 1992) showed how such periodicity coding is related to modulation information and analyzed the role of the "On" and "Chopper" neurons (in the cochlear nucleus) as preprocessors for enhancing the AM information coming from the auditory nerve fibers.

2.3 APPLICATION TO VOICED SPEECH

Many studies have concentrated on the coding of vowels (Delgutte, 1980) (Delgutte & Kiang, 1984) in the auditory nerve. For the nerve fibers whose characteristic frequency (CF) is close to a formant frequency, a phase-coupling to the formant frequency or to an adjacent harmonic is observed with little or no envelope modulation as the discharge pattern of the fiber is dominated by a single large harmonic component. Other fibers may show modulations corresponding to harmonic interactions. Therefore, the auditory system is able to track simultaneously formants and pitch by relying on phase-coupling of fibers whose CF is close to the formant ('spectral analysis') and by relying on patterns of modulation for fibers influenced by a summation of stimulus harmonics (Delgutte, 1980) (Miller & Sachs, 1984).

Most speech recognizers are based on short-term analyses that assume that the speech signal in the analysis window (typical duration of 10 to 20 ms) is stationary. As a consequence, those systems can not exploit some of the instantaneous characteristics of speech. In fact, a short-term Fourier (or linear predictive coding) analysis estimates the averaged values of harmonics on a short speech segment and can not extract accurately the patterns of modulation created by close harmonics. As these patterns seem to characterize what has been pronounced, and because they occur at glottal explosions, short-term analysis fail to exploit information that might be very useful for recognition purposes. On the other hand, most perceptive or auditory based models require a large number of inner hair cell models or cochlea filters in order to obtain a sufficiently accurate estimation of the frequency distribution of speech. To our knowledge, not much work has been done to exploit the patterns of modulation observed in the auditory nerve for formants interaction estimation and for pitch estimation.

We assume here that some of the cochlear filters and auditory nerve fibers have bandwidths broad enough to encode the envelope of the modulation produced by interacting harmonics close to formants (F1 and F2 in /a/, F2 and F3 in /i/, F3 and F4 in /i/, etc.). Therefore, the analysis does not require many cochlear filters (24) and yields an estimation of F2 - F1, F3 - F2 or F4 - F3 and of the missing fundamental by extracting the modulation pattern at the output of cochlear filters, when possible. Furthermore, there is no need to assume any stationary signal.

This system exploits the modulation information in order to perform the recognition of voiced speech segments based on patterns of modulation using a Dystal neural network (Alkon et al., 1990). We are interested in the patterns of modulation produced by 'interacting' harmonics (beats of harmonics) in medium and high frequency channels of a bank of cochlear filters in order to classify vowels and voiced sounds. The Dystal network uses the 3D speech representation delivered by the analysis to classify images that are characteristics of voiced speech.

2.4 THE ANALYSIS

The filterbank is comprised of a bank of 24 filters centered on 330Hz to 4700Hz (Moore & Glasberg, 1983)(Patterson, 1976). The output of each filter is a bandpass signal with a narrow-band spectrum centered around f_i where f_i is the central frequency (CF) of channel i. The output signal $s_i(t)$ from channel i can be considered to be modulated in amplitude and phase with a carrier frequency of f_i.

$$s_i(t) = A_i(t)\cos[\omega_i t + \phi_i(t)]$$ EQ. 2.1

$A_i(t)$ is the modulating amplitude and $\phi_i(t)$ is the modulating phase.

In this chapter we use a finite impulse response time digital Hilbert transformer (Rabiner & Schafer, 1974) to extract the envelope of $s_i(t)$. Thus

$$A_i(t) = \sqrt{s_i(t)^2 + s_i(t)_q^2}$$ EQ. 2.2

where $s_i(t)_q$ is the Hilbert transform of $s_i(t)$. Then, an image representation is obtained by plotting the product $A_i(t) \cdot A_i(t)'$ versus time and central frequency. $A_i(t)'$ is the time derivative of $A_i(t)$. The x axis is the time and the y axis is expressed in hertz according to the ERB (equivalent rectangular bandwidth) scale (Patterson, 1976). The image color is the $A_i(t) \cdot A_i(t)'$ variable.

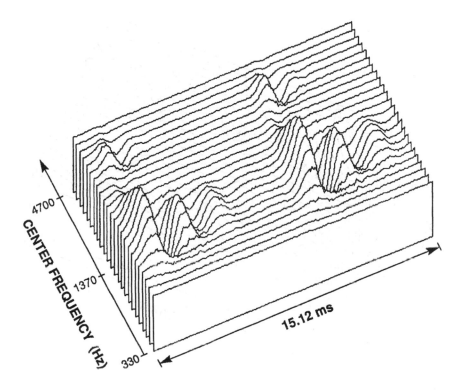

FIG. 2.1. /a/ speech segment, male speaker.

Figure 2.1 shows two pitch periods taken from a French vowel /a/ pronounced by a male speaker. The modulation occurs during glottal explosion. It is due to the interaction of harmonics close to F1 and F2 (in channels 8 to 13) and to F3 and F4 in channels 19 to 21. The period of the modulation is a multiple of T0 and is approximately equal to 1/(F2 - F1) (or to 1/(F4 - F3)) and is independent of noise power. Figure 2.2 is a three pitch periods of the French vowel /ε/. It has been pronounced by a female speaker. The modulation pattern is essentially due to F2 and F3 (in channels 17 to 19).

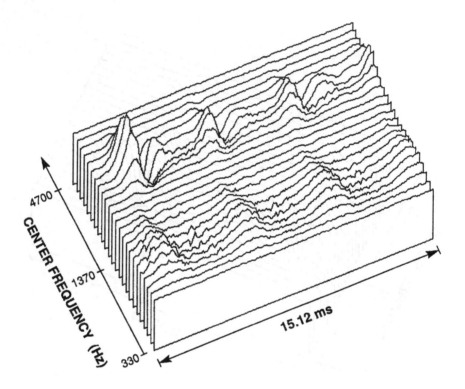

FIG. 2.2. / ε / speech segment, female speaker.

2.5 THE RECOGNIZER

2.5.1 The Dystal Neural Network

The recognizer is based on an artificial network derived from a marine snail and the hippocampus of a rabbit. The Dystal network was developed by Alkon et al. (Alkon et al., 1990). Experiments using Dystal were already reported on hand-written character recognition (Blackwell et al., 1992).

Alkon et al. (Alkon et al., 1990) proposed a network that associatively learns correlations and anticorrelations between time events occurring in presynaptic neurons. Those neurons synapse on the same element of a common postsynaptic neuron. They proposed a learning rule that modifies the cellular excitability at dendritic patches. These synaptic patches were postulated to be

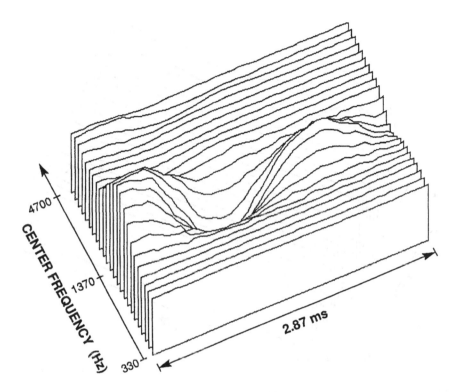

FIG. 2.3. /a/ reference pattern, male speakeı.

formed on branches of the dendritic tree of vertebrate neurons. In Dystal, weights are associated to patches rather than to incoming connection. After learning, each patch characterizes a pattern of activity on the input neurons. A Dystal network has two separate input pathways. A first pathway is the conditioned stimulus (CS) input and the second is the unconditioned stimulus (UCS). The CS input neurons are the receptive field of a Dystal neuron. The UCS inputs are used during learning and are associated to specific patterns of CS inputs.

2.5.2 The learning and the pattern recognizer architecture

The basic architecture. In training mode, Dystal learns patches corresponding to selected images (output of the multichannel analysis). The selected images are reference patterns and characterize the $A_i(t) \cdot A_i(t)'$ product for specific vowels. For example, Figures 2.3, 2.4 and 2.5 are reference patterns for the French vowels /a/, /i/ and /ε/.

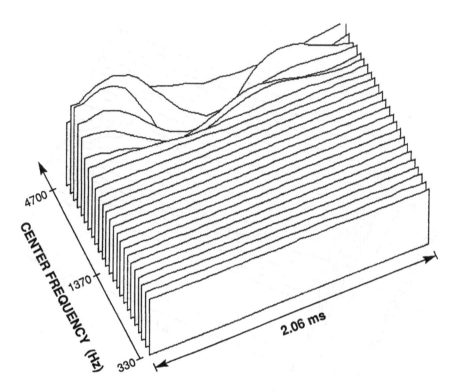

FIG. 2.4. /i/ reference pattern, male speaker.

For each patch that is being learned, an unconditioned stimulus (UCS) (a pattern name) is presented simultaneously with a conditioned stimulus (CS) (a reference pattern) . For example, the pattern of Fig. 2.3 is presented on the CS neurons input in association with the symbol /a/ as UCS input. This allows Dystal to associate the CS input pattern with the vowel /a/. For each CS image input, Dystal computes a similarity value with each of the other patterns that are already associated with the same UCS input. For example, when the UCS is /a/, Dystal computes similarities between the actual CS input and the other "a-patterns" stored in memory. If all the similarities are beneath a presettled threshold, a new pattern is created by storing the CS neurons values in patches. The patches comprise the reference pattern images and the corresponding UCS entries. If one of the similarities is above the threshold, no pattern is stored. The Dystal learning phase we use is described in (Blackwell et al., 1992).

During the recognition phase the spectro-temporal image of the spoken speech is computed. This image can be obtained in real time as there is no constraint on the duration of speech. The system generates the image for short speech segments as well as for long fluent speech sentences. A variable length window is shifted along the spectro-temporal image. For each position of the

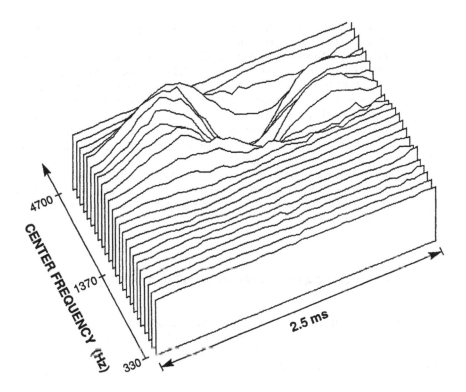

FIG. 2.5. /ℰ/ reference pattern, female speaker.

window a recognition is performed. The UCS entry of the closest reference pattern is given as output if there is a strong similarity between the reference patch and the data presented to the CS inputs. More precisely, the CS input neurons receive the data of the window with no associated UCS. The data of the window are compared with the reference patterns stored in Dystal memory. A similarity measure is computed for each class of vowels. The UCS stimulus associated with the greatest similarity is presented as output when that similarity is sufficiently high.

The two layers architecture. The two layers architecture includes the basic system as layer one. Learning and recognition are performed in the same manner and with the same conditions. In opposition to the previous architecture, no recognition decision is taken by the first layer. The output of layer one is a vector of similarity measures. The dimension of the vector is equal to the number of classes the system has to recognize. Each component of the vector is the similarity of a specific class. A vector of similarity measures is generated for each window. Consequently, the output of layer one is the time evolution of the vector of similarities. The prototype version generates a vector of similarities every 1/16 ms (i.e. the window shift is equal to 1/16 ms).

Layer two uses a fixed length window (10 ms) that is shifted (5 ms) across the image obtained by representing the time evolution of the output of layer one. After each window shift, the local maximum of the image in the window is located. A new segment that begins at the maximum location and ends 2 ms later is extracted. Similarly, for each 10 ms portion of image, a 2 ms segment is extracted following the local maximum. The segments are then concatenated. The concatenation creates a static image of similarity measures that is independent of the fundamental frequency.

A typical reference pattern for layer two consists of 10 sections of 2 ms images. The learning for layer two is exactly the same as for layer one. The difference resides in the CS inputs, that are the output images of layer one. The UCS are also the class names.

2.6 PRELIMINARY EXPERIMENTS AND RESULTS

A preliminary experiment has been conducted on four vowel classes: /a/, /i/, /y/ and / ε /. Six speakers were used for training (3 males and 3 females). A different male speaker was used for recognition. The reported results were obtained with the basic architecture.

2.6.1 Learning speakers

For each class, the reference speakers were randomly chosen. Therefore, the classes are not characterized by the same speakers. The number of reference patterns was randomly chosen. The /a/ and /i/ vowels are respectively represented by 3 reference patterns taken from 2 males and 1 female. The /y/ and / ε / vowels are also characterized by 3 reference patterns taken from 2 other females and 1 other male. The basic architecture of the system was tested with 1 male speaker in a speaker-independent mode. In the first experiment, the speaker pronounced the isolated vowels /a/, /i/, /y/ and / ε /, and the system had to recognize them. In the second experiment the speaker pronounced the isolated letters of the French alphabet.

The speech was sampled at 16000 kHz prior to be filtered by the cochlear filterbank. A speech image representation was then generated for each letter. A variable length window (2 ms to 3 ms) was placed on the spectro-temporal image. The window was shifted every 1/16 ms. For each pattern stored in Dystal, a similarity measure was computed once the window was positioned. When computing a similarity measure, the window length was automatically adapted to the length of the reference pattern under comparison.

2.6.2 Results

The recognition of /a/, /i/, /y/ and / ε / in a speaker-independent mode yielded a recognition rate of 100%. The preliminary results of the second experiment are reported below:

The vowel segments of letters a and k were recognized as /a/.

The vowel segments of letters f, l, m, n, r and s were recognized as / ε /.

The vowel segments of letters i, j, x and y were recognized as /i/.

The vowel segments of letters q and u were recognized as /y/.

The vowel segments of letters b, c and d were recognized simultaneously as /i/ and /y/.

The vowel /Φ/ from letter e was recognized simultaneously as /y/ and / ε /.

The unvoiced segments of letters f, c and s were recognized as /i/.

2.7 DISCUSSION

The preliminary results confirm that the modulation information observed during glottal explosion might be useful for classification and recognition of speech. The task presented in this paper is relatively easy and could be performed very well with standard speech recognizers (dynamic time warping, hidden Markov models, or neural networks). In fact, vowels are relatively stable and can be characterized with the distribution of averaged spectral energy. The prototype system we propose performs a very different speech analysis and exploits different information. The spectral resolution of the analysis we use is very poor. Consequently, the spectral distribution cannot be used (except for distinguishing /a/ from the 3 other vowels). But the temporal resolution is very good, and the vowels /i/, /y/ and / ε / are recognized based on the time information. This information resides in the period of the envelope of the modulated signal. It easily can be extracted and seems to be more robust to noise than the energy distribution. The training of the system is fast, once the reference patterns have been defined.

The modulation information seems to be reliable and robust to noise (Rouat et al., 1992). In comparison with standard speech recognizers (dynamic time warping, hidden Markov models, or neural networks), the system is very different. The robustness of contemporary noisy speech recognition systems is mainly due to signal processing techniques at the level of the input signal or to specific strategies embedded in the recognition process. Signal processing techniques are usually used to clean the signal and remove noise. The recognition strategies include training and learning on noisy data, estimation of the noise statistical properties, inclusion of model of noises, and so on. The robustness of the proposed system is due to the acoustical cues obtained via the modulation analysis. Therefore, the recognition process can be more simple.

Some vowels or speech classes that were never learned by the network seem to be characterized by simultaneous responses in two or more Dystal output neurons. Depending on what was pronounced, the neurons that are activated are different. Some classes that were never learned might be characterized by a strong response appearing in different output neurons. In that case, layer 2 might help to improve the potential of the approach.

2.8 CONCLUSION

We have proposed a new speech analysis, and we designed a prototype speech recognizer to evaluate the speech demodulation approach. We exploit modulation cues that are characteristics of speech and that can be used for noisy environments and auditory scene analysis. The system is based on events that are present in speech and that cannot be taken into account by the majority of contemporary speech systems.

Further experiments with many speakers and with a large vocabulary should be performed in order to evaluate the potential of that approach to speech recognition and auditory scene analysis. Furthermore, evaluations in various noisy environments have to be performed.

This chapter reports results on the exploitation of modulation information for voiced speech. More experiments have to be done in order to evaluate the pertinence of a similar approach to unvoiced speech. For unvoiced speech, the time evolution of the envelope $A_i(t)$ is a meaningful cue. It is not used by most speech systems because of the difficulty of obtaining the envelope on short segments. The processing of unvoiced speech by our system will be studied in the near future.

The prototype system has been implemented in order to study the modulation in speech and has not been yet optimized in terms of CPU time. Improvements can be done by increasing the window shift and reducing the order of the filters.

ACKNOWLEDGMENTS

This work has been supported by the National Sciences and Engineering Research Council of Canada, by the Fonds pour la Formation des Chercheurs et l'Aide à la recherche du Québec, by the Canadian Microelectronics Corporation and by the Fondation from Université du Québec à Chicoutimi. Many thanks to Daniel Morissette for his programming work.

REFERENCES

Alkon, D. L., Blackwell, K. T., Barbourg, G. S., Rigler, A. K., & Vogl, T. P. (1990). Pattern-recognition by an artificial network derived from biologic neuronal systems. *Biological Cybernetics*, 62, 363-376.

Bregman, A. S. (1984). Auditory scene analysis. *Proceedings of the seventh international conference on pattern recognition*, pp. 168-175. Silver Spring, MD: IEEE Computer Society Press.

Bregman, A. S., Abramson, J., Doehring, P., & Darwin, C. J . (1985). Spectral integration based on common amplitude modulation. *Journal of Perception and Psychophysics*, 37, 483-493.

Blackwell, K. T., Vogl, T. P., Hymans, S. D., Barbourg, G. S., & Alkon, D. L. (1992). A new approach to hand-written character recognition. *Pattern Recognition*, 25 (6), 655-666.

Delgutte, B. (1980). Representation of speech-like sounds in the discharge patterns of auditory nerve fibers. *Journal of the Acoustical Society of America*, 68, 843-857.

Delgutte, B. & Kiang, N. Y. (1984). Speech coding in the auditory nerve: Vowels in background noise. *Journal of the Acoustical Society of America*, 75, 908-918.

Gardner, R. B. & Wilson, J. P. (1979). Evidence for direction-specific channels in the processing of frequency modulation. *Journal of the Acoustical Society of America*, 66, 704-709.

Langner, G. (1992). Periodicity coding in the auditory system. *Hearing Research*, 60 (2), 115-142.

Langner, G. & Schreiner, C. E. (1988). Periodicity coding in the inferior colliculus of the cat. Neuronal mechanisms. *Journal of Neurophysiology*, 60 (6), 1799-1822.

Maragos, P., Quatieri, T. F., & Kaiser, J. F. (1992). On separating amplitude from frequency modulations using energy operators. In *Proceedings of the IEEE International Conference on Acoustics, Speech and Signal Processing*, 2.1-2.4.

Miller, M. I. & Sachs, M. B. (1984). Representation of voice pitch in discharge patterns of auditory nerve fibers. *Hearing Research*, 14, 257-279.

Moore, B. & Glasberg, B. (1983). Suggested formulae for calculating auditory-filter bandwidths and excitation patterns. *Journal of the Acoustical Society of America*, 74, 750-753.

Patterson, R. D. (1976). Auditory filter shapes derived with noise stimuli. *Journal of the Acoustical Society of America*, 59 (3), 640-654.

Rabiner, L. R. & Schafer, R. W. (1974). On the Behavior of Minimax FIR Digital Hilbert Transformers. *The Bell Systems Technical Journal*, 53 (2), 363-390.

Robles, L., Ruggero, M. A., & Rich, N. C. (1991). Two-tone distortion in the basilar membrane of the cochlea. *Nature*, 349 (6308), 413-414.

Rouat, J., Lemieux, S., & Migneault, A. (1992). A spectro temporal analysis of speech based on nonlinear operators. *Proceedings of the International Conference on Spoken Language Processing*, vol. 2, 1629-1632 edited by J. J. Ohala, T. M. Nearey, B. L. Derwing, M. M. Hodege, & G. E. Wiebe: the university of Alberta, Edmonton, Canada.

Rouat, J. (1993). Nonlinear operators for speech analysis. *Visual Representations of Speech Signals*, edited by M. Cooke, S. Beet and M. Crawford. J Wiley. pp. 335-340.

Schreiner, C. E. & Langner, G. (1988). Periodicity coding in the inferior colliculus of the cat. Topographical organization. *Journal of Neurophysiology*, 60 (6), 1823-1840.

Tansley, B. W. & Suffield, J. B. (1983). Time course of adaptation and recovery of channels selectively sensitive to frequency and amplitude modulation. *Journal of the Acoustical Society of America*, 74, 765-775.

3 A Critique of Pure Audition

Malcolm Slaney
Interval Research Corporation

All sound-separation systems based on perception assume a bottom-up or Marr-like view of the world. Sound is processed by a cochlear model, passed to an analysis system, grouped into objects, and then passed to higher-level processing systems. The information flow is strictly bottom up, with no information flowing down from higher-level expectations. Is this approach correct? In this chapter, I first summarize existing bottom-up perceptual models. Then, I examine evidence for top-down processing, describing many of the auditory and visual effects that indicate top-down information flow. I hope that this chapter generates discussion about what the role of top-down processing is, whether this information should be included in sound-separation models, and how we can build testable architectures.

3.1 THESIS

In this chapter,[1] I discuss the flow of information in a sound-analysis system. Historically, in perceptual models of audition, information has flowed from low-level filters up toward cortical or cognitive processes. The title for this chapter comes from a view that this approach, although it may offer a simple or pure way to model perception, faces increasing evidence suggesting that it is time for us to revisit this architectural model.

The question of bottom-up versus top-down processing is well known, especially in the artificial-intelligence (AI) community. In the bottom-up world, all information flows from the sensor. Bits of information are collected and combined, until finally an object is recognized. In the top-down view of the world, we know that there is a table out there somewhere; all we need to do is to collect the evidence that supports this hypothesis. Because decisions are based on sensor data, information in a top-down system flows both up and down. In real life no system lies at either extreme, but the categorization provides a useful framework to describe information flow qualitatively.

From an engineering point of view, there are many advantages to modeling the perceptual system with bottom-up information flow. As each process is studied and understood, the essential behavior is captured in a model. The results can then be passed to the next stage for further processing. Each stage of the model provides a solid footing that permits the work at the next stage to proceed.

The science of perception is bottom up. This assertion is true for both the visual system and the auditory system. Peripheral processes are studied and used as building blocks in the journey toward the cortex. It is relatively easy to understand what a neural spike near the retina or the cochlea does, but it is much harder to understand what a spike in the cortex signifies.

Churchland, Ramachandran, and Sejnowski, in their recent book chapter, "A Critique of Pure Vision" (Churchland et al., 1994), questioned the assumption that information flows exclusively bottom up. There is much evidence, both behavioral and neurophysiological, that suggests that the visual system uses significant information that flows top down. They define a *pure* system as one that is exclusively bottom up; the alternative model is a *top-down* or an *interactive* system.

We shall look at the arguments in "A Critique of Pure Vision" and shall discuss their applicability to the auditory world. Have those of us who build auditory-perception systems ignored the avalanche of information from higher cognitive levels? With gratitude to Churchland, Ramachandran, and Sejnowski, I hope that this chapter will promote discussion, and will provide a framework for describing computational auditory-scene-analysis systems.

Section 3.2 reviews the case for pure vision and pure audition systems. Section 3.3 surveys the evidence for an interactive approach to audition and vision. Section 3.4 concludes with observations about possible future research. This chapter does not describe the role of efferent projections in the auditory and visual pathways. I hope that the examples cited here will inspire more study of neural top-down connections.

3.2 MARR'S VISION

David Marr's book *Vision* was a conceptual breakthrough in the vision and AI worlds (Marr, 1982). Most important, for the present discussion, is the argument that the visual system can be described as a hierarchy of representations.[2] At the lowest level, an image represents intensity over an array of points in space. Simple processing converts these pixels into lines, curves, and simple blobs. This primal sketch can then be converted into a 2 1/2-D[3] sketch by finding orientations and noting the discontinuities in depth, all from the perspective of the camera. Later processing then converts this sketch into a world view of the objects in the scene.

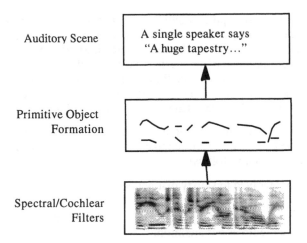

FIG. 3.1. A schematic of the pure-audition approach to auditory analysis.

Churchland and colleagues described a caricature of pure vision with the following attributes:

(1) We see the complete world. The retina records a complete image and we analyze it at our leisure.

(2) There is a hierarchy of information and representations.

(3) Information flows from bottom to top, with high-level representations depending on only the low-level processes, not vice versa.

This cartoon model of *pure vision* or *pure audition* serves as a reference point for one end of the pure–interactive scale. Whereas what Churchland called *interactive vision,* or what Blake (Blake & Yuille, 1992) called *active vision,* falls on the other end of the scale.

Many auditory systems have adapted the pure-vision philosophy. Figure 3.1 is an amalgam of system architectures as described by Mellinger (Mellinger, 1991), Cooke (Cooke, 1993), and Brown (Brown & Cooke, 1994) to do auditory sound separation. A filter stage feeds spectral information into an event analyzer and a detector. Later events are combined into objects by a process known as *scene analysis.* To my knowledge, all auditory-perception models, including my own, have assumed a bottom-up approach to the problem, often referring to Marr's notion as a guiding principle. Is this approach the best one?

Churchland, Ramachandran, and Sejnowski argued that the pure-vision view of the world is a dangerous caricature. Although computer vision has made much progress with this premise, the path could turn out to have a dead-end. They make these points:

> The idea of "pure vision" is a fiction, we suggest, that obscures some of the most important computational strategies used by the brain. Unlike some idealizations, such as "frictionless plane" or "perfect elasticity" that can be useful in achieving a core explanation, "pure vision" is a notion that impedes progress, rather like the notion of "absolute downess" or "indivisible atom" (Churchland et al., 1994, p. 24)

I worry that the same criticism applies to computational auditory scene analysis.

Churchland described—the opposite of pure vision—interactive vision or top-down processing, as follows:

(1) Perception evolved to satisfy distinct needs.

(2) We see only a portion of the visible world, although motion (or sudden sounds) can redirect our attention.

(3) Vision is interactive and predictive. It builds a model of the world and the visual system tries to predict what is interesting.

(4) Motion and vision are connected. We move to see more of the world.

(5) The neurophysiological architecture is not hierarchical; much information flows both ways.

(6) Memory and vision interact.

There is much evidence that the auditory system has many of the same properties. Perhaps our models should have them too.

A clear example of *interactive vision* is shown in Figure 3.2. Saccadic eye movements are plotted as a subject explores a visual scene. Clearly, the subject does not see the entire image at once. Instead she gradually explores pieces of it. Does a similar process occur in the auditory world?

3.3 EXAMPLES OF INTERACTIVE PROCESSES

There are many visual and auditory effects that are not what they seem. The following examples do not provide proof that the auditory and visual systems are interactive; instead, they serve to illustrate problems with a purely bottom-up view of processing flow. I describe global influences, motion, and categorization decisions that are influenced by the semantics, grouping, cross-modality influences, and the effect of learning. In all but the learning case, I give examples from the worlds of vision (from Churchland) and of audition.[4]

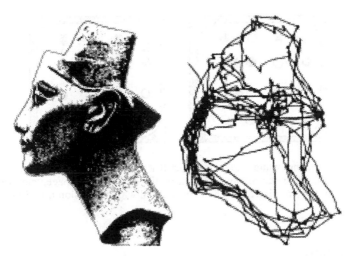

FIG. 3.2. The lines on the right are a plot the saccadic eye movements of a subject who is looking at the face on the left. (Source: Reprinted with permission from Yarbus, 1967.)

3.3.1 Global Influences

A basic feature of a pure system is that local features are all that the system needs to make decisions about the low-level properties of a stimulus. If a global property affects the local decision, then either the analysis of the two properties is different from that originally proposed, or a global or high-level information source is modifying the low-level percept.

Signal-Level Control. Both the auditory and the visual systems include control mechanisms to change the global properties of the received signal. The pupils of the visual system control the amount of light that falls on the retina. Likewise, at the lowest levels of the auditory system, efferent signals from the lower superior olivary complex affect the mechanical tuning of the cochlea, thus changing the size of the vibrations of the basilar membrane. Whereas both mechanisms are important, they do not change the information content of the signal and thus are not considered here.

Occlusion and Masking. A simple example of the type of information flow that we do want to consider is the way that we perceive occluded lines and tones. If the break is short, we see a continuous line. Likewise, if a rising chirp is partially replaced with a noise burst, we are convinced that we never heard the tone stop. The remainder of this section describes similar effects.

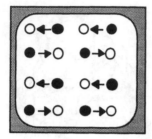 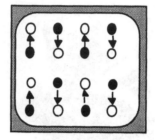

FIG. 3.3. Alternating white and black dots that create an illusion. Subjects see one uniform motion, either the motion indicated in the left or right image, and never see a combination of the two directions. (Source: Adapted with permission from Churchland et al., 1994).

3.3.2 Motion

Acoustic and visual motion provide evidence that perception is not a strictly hierarchical process. In some cases, local motion determines segmentation; in other cases, the segmentation and global properties determine the motion. A visual and an acoustic example show aspects of this hierarchy dilemma.

Vision: Bistable Quartets. Figure 3.3 illustrates a visual stimulus where the local motion is ambiguous. Motion can be perceived differently in different parts of the image; instead, however, when these two images are alternated, the subject sees all motion in the same direction. Similar examples are given in the Churchland chapter.

Audition: Deutsch Octave Illusion. Direct analogies to the bistable-quartet motion are difficult to find because acoustic-object formation is so strongly mediated by pitch and speech perception. A related auditory stimulus is presented by Diana Deutsch to show the effect of experience on perceived motion (Deutsch, 1990). Some people hear a two-tone pattern as ascending in pitch; when the pattern is changed to a different key, however, the same people hear it as descending. Deutsch reports that there is a correlation between the range of fundamental frequencies in the speaker's natural voice and the direction perceived.[5]

3.3.3 Categorization

In a purely bottom-up system, the semantic content of a stimulus does not affect the low-level perceptual qualities of a scene. Certainly, decisions such as object recognition and speech recognition are higher in the processing chain then are low-level perceptions like shape and sound characteristics.

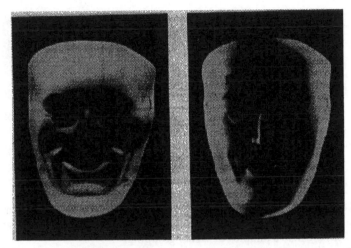

FIG. 3.4. Two concave masks photographed from their inside. The effect of faces on depth perception is illustrated. (Source: Reprinted with permission from Churchland et al., 1994.)

Vision: Faces and Shading. Figure 3.4 shows a simple example that illustrates visual ambiguity. Shading gives us important cues for determining what the shape of an object is. Most people see the masks in Figure 3.4 as having the nose projecting out of the page, even though the masks are in fact concave. Moving the lights from above—which is the direction from which we normally expect to see the light—to the sides does not change the perception that the nose is sticking out of the page as it would be from any normal face.

Audition: Ladefoged's Ambiguous Sentence. Context can dramatically affect the way speech is heard. Many people wonder whether speech is special and handled differently from other types of acoustic signals. I hope to illustrate how linguistic information and decisions can change our perception.

In Figure 3.5, the same introductory sentence is spoken by two different speakers. The final word, after each sentence, is the same—identical samples and waveforms. Yet most listeners hear the word at the end of the first sentence as "bit" and the word at the end of the second sentence as "bet." How can this difference occur if phonemes are recognized independent of their surroundings? Clearly, the words that we perceive, as shown by this example, are changed by our recent experience.

3.3.4 Grouping

Grouping together many components of a sound or scene is an efficient way for the perceptual system to handle large numbers of data. But can groups affect the low-level percepts? We would not expect a group to be formed unless all elements of the group have some property in common. Or is it possible that a

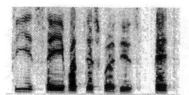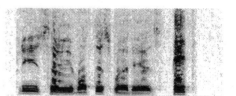

FIG. 3.5. Two spectrograms of the sentence "Please say what this word is: XX". On the left, the last word is heard as "bit," while on the right it is heard as "bet." Identical waveforms are used in both cases. (Source: Audio courtesy of Peter Ladefoged.).

high-level decision is shorthand, so many low-level decisions are unneeded? The bistable quartets in Figure 3.3, and the dots in Figure 3.7 (described in Section 3.5) are also examples of visual grouping.

Audition: Sine-Wave Speech. Speech is often described as special because we hear spoken language as words, rather than as chirps, beeps, or random noise. A large orchestra produces sounds more complicated than those made by a single vocal tract, yet it is not hard for even untrained listeners to hear the piccolo part. Yet try as we might, we have a hard time describing more of the auditory experience of speech sounds than the pitch and the loudness. Language is certainly an important grouping process.

Sine-wave speech is an example of an acoustic signal that might or might not be heard as speech (Remez & Rubin, 1993). Figure 3.6 shows a spectrogram of a sine-wave speech signal. In sine-wave speech, the pitch of the acoustic signal is removed and the formants of the speech are modeled by a small number of sine waves (three in this case).

Most listeners first hear a sine-wave speech signal as a series of tones, chirps, and blips. There is no apparent linguistic meaning. Eventually, or after prompting, all listeners unmistakably hear the words and have a hard time hearing the individual tones and blips. Some of the tones remain, but it is as though the listener's minds hear only the speech of normal speakers. With appropriate cueing, they hear the sounds as speech. The linguistic information in the signal has changed their perception of the signal.

Audition: The Wedding Song. Parts of speech can also be heard as music. Mariam Makeba recorded a musical piece called the *Wedding Song.* In the introduction, she names the song in Xhosa, an African click language. When she says the title, an American listener hears the click as part of the word. Yet when the listener hears the same type of click in the song, she hears it as separate from the speech, as part of the instrumental track. To my American-English ears, a click is not normally part of a language; when the click is placed into an ambiguous context, such as in a song, it is not heard as a part of the speech signal.

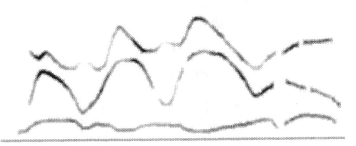

FIG. 3.6. A spectrogram of sine-wave speech. The sentence is "Where were you a year ago." (Source: Audio courtesy of Richard Remez.)

3.3.5 Cross-Modality

Thus far, we have considered the auditory and visual processing systems only independently. Surely, in a pure system, an auditory signal would not affect what we see, and visual stimuli would not affect our auditory perception. However, we do get cross-perception effects. We all perceive the voices of television actresses as coming from their mouths, rather than from the television's speakers, even if they are placed away from the screen. Two other such examples are described next.

Audition Affecting Vision: Behind the Occluder. Churchland and colleagues described a stimulus that illustrates illusory motion; it is shown in Figure 3.7. In each of three experiments, the dots in Column A are turned on and off in opposition to those in Column B (the square is always present). In the first experiment, the subject sees all three dots as moving back and forth, with the middle dot occluded by the square. In the second experiment, she sees the same dot as just blinking on and off. (These two experiments also provide an example of global changes affecting local perception.) Finally, in the third experiment, a tone is played in her left ear when the dot in Column A is shown. The dot and the left tone alternate with a tone played in her right ear. Apparent motion returns. Here an auditory event changed perception of the visual scene. How did this happen? The auditory stimuli added information to disambiguate the visual experience.

Vision Affecting Audition: The McGurk Effect. Vision can change the acoustic perception. The *McGurk effect,* an example of this cross-modality influence, is illustrated in Figure 3.8 (Cohen & Massaro, 1990). With our eyes closed, we hear a synthesized voice saying "ba." When we open our eyes, and watch the

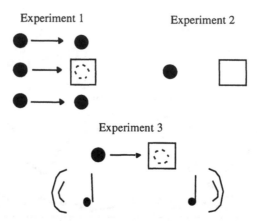

FIG. 3.7. Three experiments demonstrating illusory motion. The stimuli on the left alternate with those on the right. The arrows indicated perceived motion; the dashed circle indicates the object is perceived under the square. Global change cause the perception of motion in experiment 1; the tones played in left and right ears lead to motion in experiment 3. (Source: Adapted with permission from Churchland et al., 1994.)

artificial face, we hear "va." The acoustic signal is clearly "ba," yet the lips are making the motions for "va." Thus, our brains put together these conflicting information sources and, for this sound, trust the information from the eyes.

3.3.6 Learning

At the highest level, learning and training affect our perception over long periods. Most of the effects that we have discussed are immediate. Our perception is instantaneous and does not change much over time.

Yet training has been shown to change a owl monkey's ability to perform a discrimination task (Recazone et al., 1993). Over time, with much training, the owl monkey improved its ability to make frequency discriminations. Most important, the neurons in the AI section of monkey's cortex had reorganized themselves such that more neural machinery than before learning was dedicated to the task. A similar effect was seen with visual discrimination

3.4 FUTURE WORK

I know of no study that quantifies the information flow down the processing chain. Clearly, the centrifugal or descending auditory pathways are important. At the lowest levels, efferent signals from the superior olivary complex affect the mechanical tuning of the cochlea.

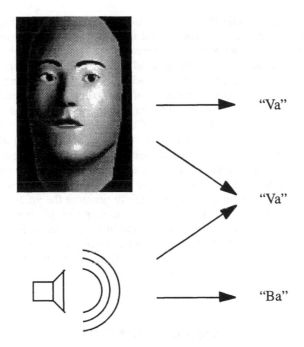

FIG. 3.8. The McGurk effect illustrates how visual stimuli can overrule the auditory perception. (Source: Image courtesy of Michael Cohen, University of California, Santa Cruz.)

Many of the examples in this and Churchland's chapter can be explained easily if low-level detectors generate all possible hypothesis. Higher-level processes then evaluate all the ideas, and suppress the inconsistent results. It is impossible for psychophysical experiments to rule out one or the other of these alternatives. To answer this question, we must perform experiments on efferent projections.

There are auditory systems that use top-down information. Most speech-recognition systems today use linguistic information and knowledge about the domain to guide the word-hypothesis search (Lee, 1989). A system proposed by Varga and Moore (1990) uses two hidden Markov model (HMM) recognizers to separate speech and noise. Works by Carver and Lessor (1992), Nawab (1992) and Ellis (1993) discuss blackboard systems that allow expectations to control the perception. Recent work by the Sheffield group (Cooke et al., 1995) used Kohonen nets and HMMs to recognize speech with missing information. These systems, however, are not tied to physiology or psychoacoustics. Is there common ground between speech recognizers and systems that perform auditory-scene analysis?

I do not mean to imply that pure audition is inherently bad. Interactive and top-down systems are hard to design and test. The world of perception offers little guidance in the design of these systems.

Instead, I hope to find a middle ground. I hope that those of us who design top-down systems will learn what has made the perceptual system successful. We wish to discover which attributes of the perceptual representation are important and should be incorporated into the top-down systems. Clearly the mel-frequency cepstral-coefficient (MFCC) representation in the speech-recognition world (Hunt et al. 1980) is one such win for perception science.

Likewise, those of us who design pure-audition systems need to acknowledge all the top-down information that we are ignoring in the pursuit of our sound-understanding systems. Much information is processed without regard to high-level representations. We clearly perceive the voice of somebody speaking a language we have never heard as being one sound source, rather than as isolated chirps and tones. Yet many problems—such as understanding how we separate speech from background at a noisy cocktail party—might be easier to solve if we pay attention to our understanding of the linguistic content. I, unfortunately, do not know how to do so yet.

ACKNOWLEDGMENTS

Earlier versions of this work were presented to the Perception Group at Interval Research and to the Stanford CCRMA Hearing Seminar. I am grateful for the feedback, criticism, and suggestions I received from all these people. Specifically, Subutai Ahmad, Michele Covell, and Lyn Dupré had many suggestions that greatly improved my ideas and their presentation.

REFERENCES

Blake, A. & Yuille, A. (Eds.). (1992). *Active Vision*. Cambridge, MA: MIT Press.

Brown, J. G. & Cooke, M. P. (1994). Computational auditory scene analysis. *Computer Speech and Language*, 8 (4), 297–336.

Carver, N. & Lessor, V. (1992). Blackboard systems for knowledge-based signal understanding. In *Symbolic and Knowledge-based Signal Processing*, Alan V. Oppenheim and S. Hamid Nawab, (Eds.). Englewood Cliffs, NJ: Prentice-Hall.

Churchland, P., Ramachandran, V. S., & Sejnowski, P. (1994). A critique of pure vision. In *Large-Scale Neuronal Theories of the Brain,* Christof Koch and Joel Davis, (Eds.). Cambridge, MA: MIT Press.

Cohen, M. M. & Massaro, D. (1990) Synthesis of visible speech. *Behavior Research Methods, Instruments and Computers*, 22(2), 260–263.

Cooke, M. (1993). *Modelling Auditory Processing and Organisation*. Cambridge, UK: Cambridge University Press.

Cooke, M., Crawford, M., & Green, P. (1995). Learning to recognise speech in noisy environments. ATR TR-H-121, *Proceedings of the ATR workshop on A Biological Framework for Speech Perception and Production,* Kyoto Japan, 13–17.

Deutsch, D. (1990). A link between music perception and speech production. *Abstracts for the One Hundred Twentieth Meeting of the Acoustical Society of America, Journal of the Acoustical Society of America*, Vol. 88 (Suppl. 1).

Ellis, D. P. W. (1993). Hierarchic models of hearing for sound separation and reconstruction. *1993 IEEE Workshop on Applications of Signal Processing to Audio and Acoustics*, IEEE, New York, pp. 157–160.

Hunt, M. J., Lennig, M., & Mermelstein, P. (1980). Experiments in syllable-based recognition of continuous speech. *Proceedings of the 1980 ICASSP*, Denver, CO, pp. 880–883.

Ladefoged, P. (1989). A note on 'information conveyed by vowels.' *Journal of the Acoustical Society of America*, 85 (5), 2223–2234.

Lee, K. (1989). *Automatic Speech Recognition: The Development of the SPHINX System*. Boston, MA: Kluwer Academic Publishers.

Marr, D. (1982). *Vision*. San Francisco, CA: W. H. Freeman and Company.

Mellinger, D. K. (1991). Event formation and separation in musical sound. Doctoral Thesis, CCRMA, Dept. of Music, Stanford University, Stanford, CA.

Nawab, S. H. (1992). Integrated processing and understanding of signals. In *Symbolic and Knowledge-Based Signal Processing*, Alan V. Oppenheim and S. Hamid Nawab, (Eds.), Englewood Cliffs, NJ: Prentice-Hall.

Recazone, G. H., Schreiner, C. E., Merzenich, M. M., (1993). Representation of primary auditory cortex following discrimination training in adult owl monkeys. *The Journal of Neuroscience*, 13(1), 87–103.

Remez, R. E. & Rubin, P. E. (1993). On the intonation of sinusoidal sentences: Contour and pitch height. *Journal of the Acoustical Society of America*, 94 (4), 1983–1988.

Varga, A. P. & Moore, R. K. (1990). Hidden Markov model decomposition of speech and noise. In *Proceedings of ICASSP-90*, Albuquerque, NM, Vol. 2, 845–848.

Yarbus, A. L. (1967). *Eye Movements and Vision*. New York: Plenum Press.

NOTES

[1]An earlier version of this chapter was published as Interval Technical Report IRC1995-010.

[2]A second important aspect of Marr's work deals with the representations of the "information processing" task. A *computational learning theory* specifies what the goal of an algorithm is and why it is important. The *representation* and *algorithm* specify how the computational theory should be implemented. Finally, the *hardware representation* describes how the algorithm is realized. These distinctions are important in both the audition and vision worlds, and should be kept clearly in mind.

[3]Marr's book (Marr, 1982, pp. 128–129) describes a 2 1/2-D sketch as follows. "According to our emerging theory of intermediate visual information processing, however, a key goal of early visual processing is the construction of something like an orientation-and-depth map of the visible surfaces around a viewer. In this map, information is combined from a number of different and probably independent processes that interpret disparity, motion, shading, texture, and contour information. These ideas are called the 2 1/2-D sketch. ... The full 2 1/2-dimensional sketch would include rough distances to the surfaces as well as their orientations; contours where surface orientation change sharply, which are shown dotted; and contours where depth is discontinuous (subjective contours), which are shown with full lines."

[4] Many of the examples in this chapter can be found online at
     http://www.interval.com/papers/1997-056

[5] Furthermore, there is a strong difference in perception between subjects who grew up in California and those who grew up in the South of England.

4

Psychophysically Faithful Methods for Extracting Pitch

Ray Meddis and Lowel O'Mard
Loughborough University of Technology, UK

Pitch is a potentially critical component of an auditory scene analysis system. It is therefore important that the pitch extraction system is robust and flexible. An algorithm is outlined which is faithful to psychophysical phenomena. It has been successfully used to segregate simultaneous vowels by emphasizing frequency-selective channels on the basis of pitch signatures. Two recent challenges to the pitch extraction algorithm are addressed. These suggest that there are at least two different pitch segregation mechanisms; one is accurate, phase insensitive and uses low harmonics; the other is inaccurate, is phase sensitive and uses high harmonics. Explorations using the model show that it is able to account for these new data without amendment. The second challenge involves a suggestion that pitch may normally be used to exclude channels rather than select them is accepted as plausible.

4.1 INTRODUCTION

Auditory scene analysis (Bregman, 1990) offers a general approach to the problem of extracting information concerning a sound source embedded in a noisy auditory background. The idea is to use some known physical characteristic of the sound source to isolate the relevant information from its background and thus assist with its analysis. These physical characteristics might include time of occurrence, pitch, formant trajectory, similarity to some known and expected event, etc.. This process is often treated as equivalent to the psychological phenomenon of attention. Although the strategy is clear, the tactics remain to be decided. This chapter concerns two possible tactics; channel selection and the use of voice-pitch in selecting channels.

The channel selection strategy requires a bank of bandpass auditory filters modeled on what we know of the mammalian auditory periphery. The process yields a number of parallel representations of the signal known *as frequency-selective channels* or *channels* for short. When the input to the system consists of a signal embedded in a complex background of interfering sounds, we can make the reasonable assumption that the spectrum of the foreground and background sounds are different, in which case, some of our channels will be dominated by the foreground and some by the background sound source. If we can determine which channels are dominated by the target sound source, we may seek to enhance our signal to noise ratio by selecting (or merely emphasizing) those channels which are dominated by the target signal (foreground sound source). Conversely, we may reject or attenuate the output of those channels that are dominated by the background (non target) sound source.

Even if we know which channels to select, there is no guarantee that this is the optimum stratagem. It is simply one of a number of approaches open to the auditory scene analyst. However, it certainly does work under certain circumstances. We showed that it can be used to enhance the identification of simultaneous vowels when the two vowels have a different pitch (Meddis & Hewitt, 1992). Brown and Cooke (1995) also explored this technique with the added benefit of reconstructed sounds based on selected channels.

There is a range of possible methods for selecting the channels. One method might be top down in the sense that the higher levels of analysis "know" which are the best channels to select, perhaps on the basis of past experience in similar situations. Our preference is to study low-level methods that use simple acoustic characteristics of the sound such as onset time or, in the case of harmonic stimuli, the fundamental frequency or pitch.

In principle, at least, it is not too difficult to identify onset events within a channel, and this will permit us to apply a common emphasis to channels with a simultaneous onset. This method has the advantage that it does not require any particular kind of structured sound source; it will work equally well with noise burst as meaningful speech. It assumes only that the spectral components have reasonably synchronous onsets and tolerably short attack times. On the other hand, it is difficult to know how to assign an onset to a particular sound source and the system could be prone to switch indiscriminately between emphasizing foreground and background channels.

Pitch has the advantage of being characteristic of a sound source; that is, two sound sources may have different pitches. Knowing which pitch to follow may allow the target sound to be tracked over a long period of time so long as the pitch does not vary too much and so long as the interfering sound retains a different pitch or no pitch at all. The process of extracting pitch and identifying channels with a common pitch is, however, nontrivial.

The problem is illustrated in Figure 4.1, which shows the output of a filter bank when driven by a set of 30 harmonics of a 100 Hz fundamental. All channels are driven by the same sound source but this is not immediately obvious from the figure. In the high harmonic region each channel is pulsing at

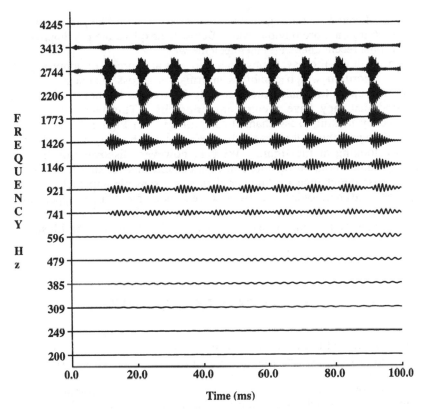

FIG. 4.1. Selected filter outputs for a harmonic complex, with 30 harmonics and a 100 Hz fundamental frequency.

a 100 Hz rate and the grouping problem is relatively trivial. This beating occurs because more than one harmonic is passing through each filter and the output pulses at the difference frequency of adjacent harmonics. This rate is the fundamental of the complex. However, in the low harmonic region individual frequencies are resolved, that is, each channel is dominated by only one frequency. Here, the pulsing is less obvious, and the basis for assigning them to the same group is not immediately clear.

Psychophysical studies (Ritsma, 1967) tell us that human listeners report a much clearer pitch percept when presented with an isolated group of low harmonics than they do when presented with an isolated group of high harmonics. This is the opposite of what we would expect from a visual examination of Figure 4.1. This tension between "the obvious way to do it" and the psychophysical reality has been the basis of a long-running century-long controversy concerning the nature of human pitch perception (Moore, 1989). This represents a major challenge to auditory scene analysis if pitch is to be used as the basis of sound source segregation. Do we try to do it like people do it or do we try do it in a more obvious way? Below we explore the psychophysical route and some recent challenges met upon the way.

If we go down the psychophysical road, we are forced to adopt a pitch extraction algorithm which combines information across channels. This is because a group of low harmonics can only be seen as harmonic when it is appreciated that they have a common periodicity. This fact can only be appreciated by making cross-channel comparisons. In Figure 4.1 all of the components repeat themselves every 10 msec. Again, this may not be obvious merely by looking at the figure. We can, however, discover this fact by computing autocorrelation functions in each channel and looking for common periods across the channels. Autocorrelation is a sensible approach for both high (unresolved) harmonics and low (resolved) harmonics. It so happens that, it yields a more precise estimate when using low harmonics and this is fortunately consistent with the psychophysics. This point will be illustrated below.

4.2 A PITCH EXTRACTION ALGORITHM

In an explicit attempt to model the psychophysical approach, we have used the four stage model of pitch perception shown in Figure 4.2. This consists of a bank of band pass filters followed by a bank of inner hair cell models that provide the input to an autocorrelation stage. The process is completed by a summation process which adds together all the autocorrelation functions (ACF). The pitch of the signal is estimated using the reciprocal of the lag of the highest peak in the summary autocorrelation function (SACF). Full details of the computational procedures are given elsewhere (Meddis & Hewitt, 1991).

Figure 4.3 shows the resulting ACFs and the SACF for a stimulus consisting of the first 30 harmonics of a 100 Hz fundamental. The major peak in the SACF occurs at a lag of 10 msec. This represents the model's best estimate (100 Hz = 1/10 msec) of the fundamental frequency of the complex.

The role of the inner hair cells is not immediately obvious but it is essential to the functioning of the system. They act as low-pass filters and this results in a reduction in the heights of the peaks in the ACF of the high frequency channels. This can be seen in the top half of Figure 4.3. Ultimately, this reduces the contribution of the high frequency channels to the pitch estimate; a property which is in line with the known psychophysics.

The SACF reflects an average of all of the activity in the ACFs. Because the different channels are responding at different frequencies, the summation process leads to a 'canceling out' effect along much of its length for a harmonic stimulus. However, all ACFs of such stimuli have peaks at the lag corresponding to the fundamental frequency. In Figure 4.1 this can be seen as a

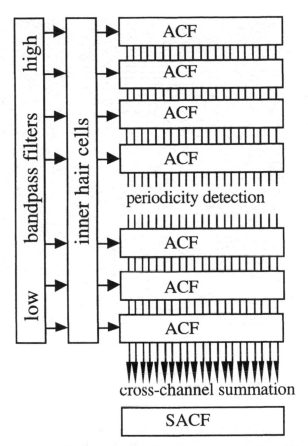

FIG. 4.2. Schematic diagram of a four-stage pitch extraction method. The pitch is taken as the largest peak in the summary autocorrelation function (SACF).

ridge up the ACF surface at a 10 msec lag. At this lag the SACF does not cancel out but accumulates across channels and a major "pitch peak" arises.

We found (Meddis & Hewitt, 1991) that it was possible to predict many psychophysical pitch matching results using this model. These included; pitch of the missing fundamental, ambiguous pitch, pitch shift of equally-spaced enharmonic components, musical chords, repetition pitch, the pitch of interrupted noise, the existence region and the dominance region for pitch. The model was also indicative of a number of aspects of listeners' sensitivity to the phase relationships among harmonic components of tone complexes. The phase sensitivity was successfully evaluated using i) amplitude-modulated and quasi frequency-modulated stimuli, ii) harmonic complexes with alternating phase and monotonic phase change across harmonic components and iii) phase effects associated with mistuned harmonics.

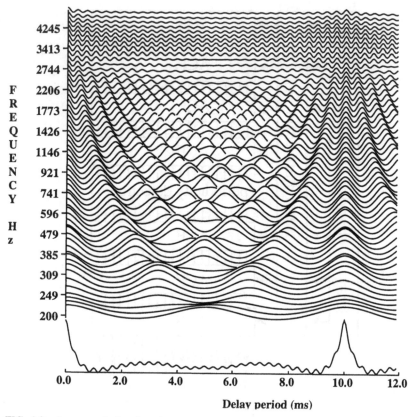

FIG. 4.3. Autocorrelation functions and summary autocorrelation functions produced by the model shown in Figure 4.2. The input is a 30 harmonic complex with a fundamental frequency of 100 Hz.

4.3 VOWEL SEGREGATION

It seemed likely that this method of pitch extraction was potentially useful in an auditory scene analysis context. To test this we used a double vowel paradigm where the two vowels could have either the same or a different pitch. When the two vowels had different pitches, two peaks were observed in the SACF. When we chose one of these peaks as characteristic of the sound source of interest we were able to define a group of channels that had peaks in their ACF corresponding to the peak in the SACF. By eliminating other channels from consideration, we sought to identify the vowel on the basis of the remaining

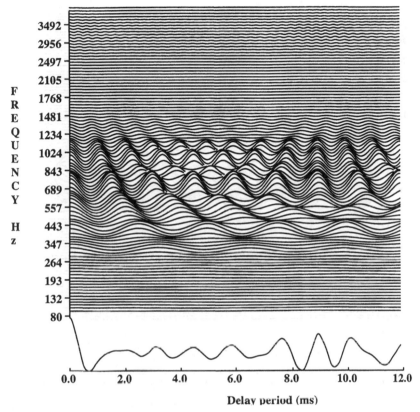

FIG. 4.4a. Model output using a stimulus consisting of two simultaneous synthesized vowels "er" and "ah" with F0 of 100 and 112 Hz respectively. Note the two the "pitch peaks" at 10 and 9 msec.

   a. unadjusted output.

   b. output after removing channels which do not have a common pitch peak at 10 ms.

   c. the remaining channels; that is, the complement of b.

The identification of the segregated vowels are based on b and c respectively.

channels. We then repeated the operation using the remaining channels to identify the second vowel. Our investigations (Meddis & Hewitt, 1992) established that the system worked and could be used to identify double vowels more accurately when a pitch difference existed.

    Figure 4.4 illustrates the process. The first image shows the initial response of the model to a stimulus consisting of two simultaneous, synthesized vowels starting simultaneously and ending after 200 msec. One vowel ("er") had a fundamental frequency of 100 Hz while the other vowel ("ah") had a fundamental frequency of 112 Hz. Figure 4.4a shows the state of the model just as the stimulus is about to cease. The SACF is more confused than in Figure 4.3

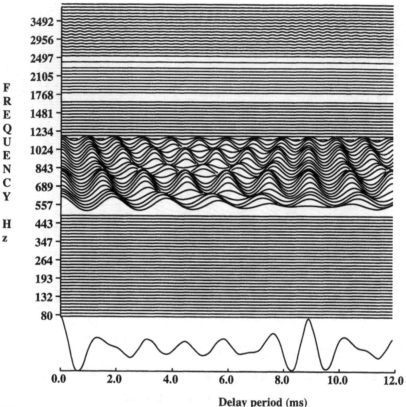

FIG. 4.4b.

but does have two pitch peaks at lags of 10 and 9 msec. We can use the larger of these two peaks (9 msec.) to group the channels into two subsets. In Figure 4.4b we selected all the channels with peaks at 9 msec lags in their ACFs. This represents the set of channels presumed to be dominated by a sound source with a pitch of 112 Hz. In Figure 4.4c we used the remaining channels. This represents the set of channels presumed to be dominated by a sound source with a higher pitch of 112 Hz.

In this case, the term 'channel selection' can be interpreted as 'channel rejection'. The output of the rejected channels is completely attenuated in Figures 4b and 4c. However, we found that the results were not greatly affected if we merely attenuated the "rejected" channels by 50%. The system is similar to a graphic equalizer which is controlled by pitch indicators. It provides a primitive differentiation of foreground and background. Attenuated channels constitute the background while the remainder constitute the foreground. Human listeners can switch between foreground and background under

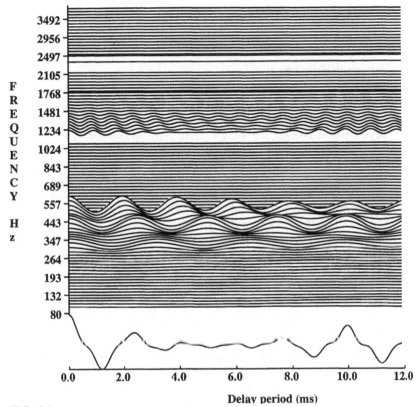

FIG. 4.4c.

conscious control. However, we propose that this control is only possible because of the low-level pre-processing afforded by the system.

The vowel recognition algorithm used the left hand portion of the SACF (from 0 to 4.5 msec lags) of isolated vowels to form templates that were stored. These were then matched to the observed SACFs for test stimuli. It can be seen that this region of the SACF is different for Figures 4a and 4b. This was the basis for the identification of the vowels. We were not studying recognition algorithms specifically at the time and this method was used *faute de mieux* but it was useful and may repay further evaluation in its own right.

4.4 CHALLENGES TO THE MODEL

4.4.1 Channel Selection

Recently, there were two direct challenges to our approach. The first challenge came from Andrew Lea (1992) who showed that double vowel identification was preserved when one of the two vowels was whispered (i.e., was not harmonically structured). In this event, only one peak will appear in the SACF and, according to our system, only one vowel could be isolated and identified.

Lea suggested that the channel selection system was simpler than we had proposed. He argued that one of the two vowels would normally be dominant and only lightly masked by the second vowel. No signal enhancement would be required to identify that vowel. On the other hand elimination of the channels dominated by that vowel would be necessary to permit the identification of the weaker vowel. He proposed that the system first identifies the dominant vowel without any special assistance. It then attenuates the output from channels with the dominant pitch signature in order to better identify the weaker stimulus.

We reprogrammed our system using this more elegant approach and recalculated our results using our original double vowel stimuli. The results (unpublished) were very similar to those obtained earlier and we, therefore, recommend consideration of Lea's method when implementing a pitch-based, channel-selection strategy for segregating harmonic sounds from their backgrounds. Using the example in Figure 4.4, this means that we should base our estimate of the first vowel directly on the untreated SACF in Figure 4.4a. The second (nondominant) vowel should then be estimated using the SACF in Figure 4.3c; that is, after the elimination of ACFs that have a peak corresponding to the pitch of the dominant vowel.

This approach leads to an interesting twist. It appears to suggest that the "cocktail party" phenomenon is not an enhancement of proximal sounds (the person you are talking to) but an enhancement of masked sounds (the person you are not supposed to be talking to).

4.4.2 One or Two Mechanisms?

A second challenge to our pitch extraction mechanism arose in the context of a debate among psychoacousticians as to whether there is one mechanism or two for extracting pitch. Houtsma and Smirzynski (1990) showed that the pitch of isolated groups of harmonics was accurate for low harmonics (JND about 0.2%) but less accurate for groups of high harmonics (JND about 2%); that is, an order of magnitude different. Moreover the transition from accurate to inaccurate occurred quickly at around 2 kHz when using a 200 Hz fundamental.

A. sine phase

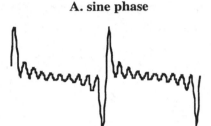

B. alternating phase

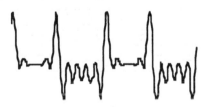

FIG. 4.5. The effect of alternating phase on two cycles of a 10 harmonic stimulus with 100 Hz fundamental.

More recently, Shackleton and Carlyon (1994) showed that alternating the phase of successive harmonics can produce a doubling of the perceived pitch of the tone complex when using a group of high harmonics but not when using a group of low harmonics. Again, this suggested the presence of two complementary mechanisms for extracting pitch, one (inaccurate and phase sensitive) for high harmonics and another (accurate and phase insensitive) for low harmonics.

Figure 4.5. shows the stimulus waveform for a 10 harmonic stimulus Figure 4.5a shows a stimulus consisting of components in sine phase. Figure 4.5b shows a stimulus with the same components but in alternating (sine, cosine, sine...) phase. It is clear that the alternating phase stimulus has an envelope with half the period of the sine phase stimulus. If the periodicity of the stimulus determines the perceived pitch, we must expect a doubling of the pitch with alternating phase. On the other hand, if the individual harmonic component frequencies are used as the basis, we must expect no change in pitch because the component frequencies have not changed. Shackleton and Carlyon's results, therefore, suggest that human listeners are sensitive to the envelope of the

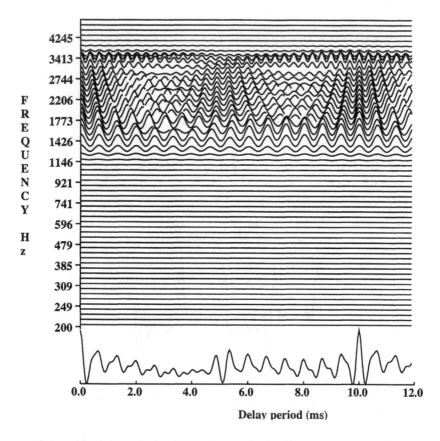

FIG. 4.6a. Model response to an alternating phase harmonic signal (F0 100 Hz). a.
Using harmonic numbers 25 - 40. b. Using harmonic numbers 1 - 10.

stimulus when high harmonics are used but not when low harmonics are used;
that is, two separate pitch mechanisms may be at work.

This challenge to our single mechanism model is blunted somewhat by our
earlier observation (see Figure 4.1) that filter outputs for resolved (low
numbered) harmonics and unresolved (high numbered harmonics) offer quite
different visual clues to the pitch of the stimulus. Despite this, the model is
quite successful in predicting the results of many pitch studies. It seemed
possible, therefore, that our model would survive this challenge without
amendment.

Figure 4.6 illustrates the response of the model to groups of a) high and b) low
harmonics presented in alternating phase. The SACF for the low harmonics
looks very similar to that in Figure 4.3 in that it shows a distinct single peak at a
lag of 10 msec. For high harmonics, however, the response is quite different
and shows a new peak at a lag of 5 msec corresponding to a prediction of a

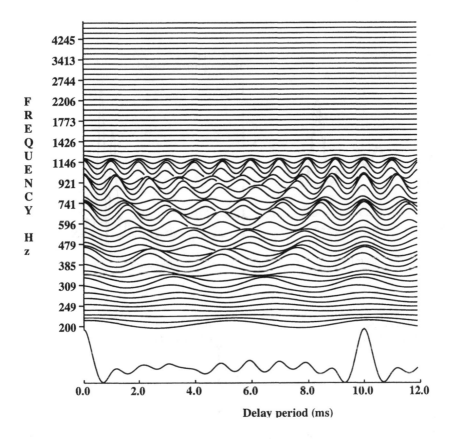

FIG. 4.6b.

doubling of the pitch. The new peak at half the fundamental period is not quite as large as the peak at the 10 msec fundamental period. It does, however, represent a substantial change from the low harmonic stimulus which has a trough at the half period point. We may need to modify our rule slightly concerning the choice of the highest peak as the exclusive basis of pitch judgments. Nevertheless, it is clear that the model is responding appropriately to alternating phase stimuli and is not invalidated by this new psychophysical demonstration. The next problem is that of the relative accuracy of pitch judgments based on low and high harmonic stimuli. We studied this problem of sensitivity to pitch change by computing the SACF for a harmonic complex with F0 of 100 Hz and 102 Hz; that is, a 2% shift in the fundamental frequency. The SACFs for the two stimuli are superimposed on one another to the left of Figure 4.7. This was repeated for groups of low harmonics (top row) and high harmonics (bottom row).

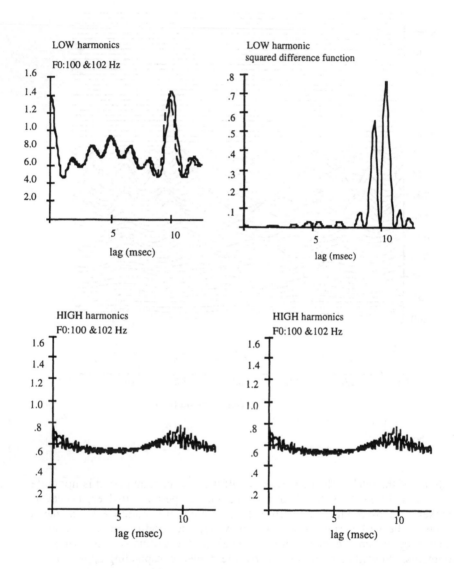

FIG. 4.7. SACFs and difference functions for LOW and HIGH harmonic stimuli. The function on the right is the squared difference between the SACF for 100 Hz and the SACF for 102 Hz. The scales are arbitrary but comparable across figures.

The stimulus consisting of low harmonics (200- 600 Hz) gives a very clear pitch peak while the high harmonic (3500- 4700 Hz) stimulus yield a much less distinct hump whose central point is difficult to locate. This is consistent with listeners' reports that the pitch percept is very clear for low harmonics but increasingly indistinct as the number of the lowest harmonic is raised.

On the left of the figure we show the difference of the two SACFs The difference function is our attempt to represent the basis on which a judgment would be made as to whether a pitch shift had occurred between the two stimuli (100 and 102 Hz). The peak of the difference function for the low (resolved) harmonics is substantially higher than for the high (unresolved) harmonics. This suggests that a 2% pitch shift will be much more noticeable for low harmonic stimuli. Qualitatively therefore, the result is in the right direction; high accuracy for low harmonics and low accuracy for high harmonics.

On the basis of these two studies it seems likely that there is no need to postulate two separate mechanisms to explain the psychophysics data. Our original model can account for them without amendment

4.5 CONCLUSIONS

The four-stage pitch extraction algorithm now has a long list of psychophysical phenomena that it can simulate. As such it is an obvious candidate for use in auditory scene analysis programs. It will certainly provide an intellectually useful bridge between computational approaches to auditory scene analysis and psychophysical science. The four stage approach is, however, computationally expensive and the benefits of a psychophysically faithful simulation will need to be made explicit to justify the extra effort. The search for indisputable benefits will undoubtedly dominate the next phase of this research.

Channel selection techniques will also require further investigation. They have the advantage that they may well map directly onto known physiological structures which store information tonotopically and involve inhibitory connections between tonotopic regions. Our aim is to show that the "channels" in our model are the same as these tonotopic regions and to develop a model which holds the psychology, psychophysics and physiology within a mutually informative theoretical framework.

ACKNOWLEDGMENTS

This research was supported by a grant from the Image Interpretation Initiative of the Science and Engineering Research Council, UK (GR/H52634).

REFERENCES

Bregman, A. (1990). *Auditory Scene Analysis - The Perceptual Organisation of Sound.* Cambridge, MA: MIT Press.

Brown, G. & Cook, M. (1994). Computational auditory scene analysis. *Computer Speech and Language*, 8, 297 -336.

Houtsma, A. & J. Smirzynski, J. (1990). Pitch identification and discrimination for complex tones with many harmonics. *Journal of the Acoustical Society of America*, 87, 304 - 310.

Lea, A. P. (1992). *Auditory Modelling of Vowel Perception*. Ph.D. thesis, Sheffield University, UK.

Meddis, R. & Hewitt, M. (1991). Virtual pitch and phase sensitivity of a computer model of the auditory periphery. I: Pitch identification. *Journal of the Acoustical Society of America*, 89, 2866 - 2882.

Meddis, R. & Hewitt, M. (1992). Modelling the identification of concurrent vowels with different fundamental frequencies, *Journal of the Acoustical Society of America*, 91, 233 - 245.

Moore, B. C. J. (1989). *An Introduction to the Psychology of Hearing*. London: Academic Press.

Ritsma, R. J. (1967). Frequencies dominant in the perception of the pitch of complex sounds, *Journal of the Acoustical Society of America*, 42, 191-198.

5

Implications of Physiological Mechanisms of Amplitude Modulation Processing for Modeling Complex Sounds Analysis and Separation

Frédéric Berthommier
Institut de la Communication Parlée

Christian Lorenzi
PCH, Institut de Psychologie

We propose a model of amplitude modulation processing and we compare its performances with psychoacoustical data. Sensitivity to modulation depth, masking between two modulations and sensitivity to their dephasing are investigated. Mechanisms of amplitude modulation (AM) processing are discussed in light of these experiments. We conclude that there are other plausible neural mechanisms than the autocorrelogram for explaining AM sensitivity and harmonic grouping in the intermediate levels of the auditory system.

5.1 INTRODUCTION

Complex sound separation requires an intermediate processing level between the peripheral stage and the identification stage, with at least two channels in this intermediate level (Berthommier et al., 1989), namely a *phasic* one coding for onsets and a *tonic* one coding for frequency content. We propose a speech processing model based on a tonic/phasic architecture (Schwartz et al., 1992).

These channels roughly correspond to cells populations in the cochlear nucleus responding to events and periodicity. Psychophysical and modeled properties of grouping and segregation of auditory objects depend on the representation of such features (Bregman, 1990, Cooke et al., 1993). The older model of *periodicity analysis* is the Licklider's autocorrelogram, used by Assmann and Summerfield (1990) for vowels segregation. But no physiological experiment has shown evidence in favor of such direct autocorrelation processing of the periodicity in the auditory system.

5.2 BASIC MODELING

Our modeling study is based on a probabilistic stellate cell model representing the activity of "integrate and fire" units (Berthommier, 1992), or uses the Hodgkin-Huxley system (HH) (Lorenzi & Berthommier, 1994). It is comparable with what is proposed by Hewitt et al. (1992), but it establishes a clear link among these three kinds of implementations. This gives at the same time a powerful computational tool for physiological modeling.

5.2.1 Cochlear Nucleus Level

The number of parameters of our probabilistic model is small: threshold, time constant of the dendritic filtering, absolute refractory period (PRA, in French), time constant of the relative refractory period (PRR, in French), and the computational complexity is low.

The principle for obtaining the non-linear characteristics of stellate cells is an association between a threshold and two filters, one forward and one backward, respectively corresponding to dendritic and refractory processes. It is remarkable to see that the probabilistic absolute refractory process is well defined by a delayed differential equation. The use of *delay lines* is a great similarity with an autocorrelation method. But the basic refractory function performs a product between the current input and the probability R(t) to be outside the refractory period (i.e. to have no previous discharge during the PRA delay), itself depending on delayed outputs, whereas autocorrelation results from a product between current and delayed inputs. In contrast with the autocorrelogram, long delay lines needed for low modulation frequency processing are biologically plausible if they correspond to a recovery dynamic instead of a propagation process.

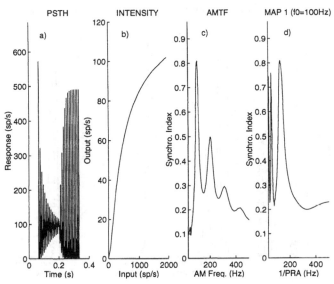

FIG. 5.1. Resonance and transfer function of stellate cells at the cochlear nucleus level. This is computed with the probabilistic model described in text. The mean input rate is a sampled signal (10KHz) corresponding to a Poisson point process. The mean discharge rate is always 1500 spikes/sec. For an AM/rectified signal, the modulation depth is 0.5. Parameters of the model: threshold = 4, dendritic filter = 1ms, PRA = 8ms, PRR = 6ms. a) PSTH of this cell for a constant signal followed by an AM signal modulated at 100Hz, near the BMF of the cell. We show that the modulation depth is greatly increased. b) Saturation of the mean discharge rate. Input is a constant signal. This model has a weak threshold effect which can be easily increased. c) Amplification of the synchrony index (input = 0.25) relative to frequency, showing a best frequency and a little near-harmonic rebound. Signal is filtered for other AM frequencies. d) Synchrony index on the AM frequency of the stimulus (100Hz), computed for an array of stellate cells. Hence, this is relative to the driving signal. The variable is the absolute refractory parameter PRA. Other parameters remain the same. Cells with PRA = 8ms are better synchronized. Secondary peaks are observed for very long (upper to 10 ms) refractory periods, probably not realistic for stellate cells. This resonance of the cells is suitable for building a rate place map at the inferior colliculus level.

When the driving signal has a period compatible with the PRA duration, the probability R(t) periodically increases in synchrony with the input, so that resultant outputs are well synchronized on the input frequency. The phase locking property is attested by the amplitude modulation transfer function (AMTF), showing multiple peaks, one for the best modulation frequency (BMF) and another one well tuned to the BMF first harmonic (Figure 5.1). We must mention that the model is tuned here in order to obtain such great modulation gains. This is motivated by functional constraints and not by the intention of well fitting known stellate cells responses.

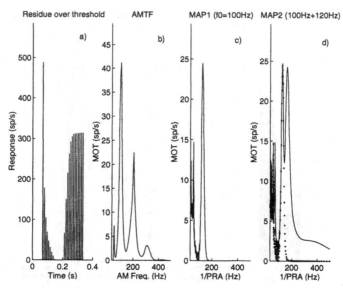

FIG. 5.2. AM frequency mapping at the inferior colliculus level. Responses are computed with the output of a probabilistic model with varying absolute refractory periods (PRA) giving a correlated variation of BMF. Threshold values of IC cells are given by a first order low pass filter (time cst = 5ms) multiplied by a constant (= 1.45). a) Residue of excitation with AM frequency = 100Hz. This residue is integrated over the second (periodic) part of the excitation. The mean discharge rate is the MOT (Mean over threshold) corresponding to the activity of physiological IC cells. b) Transfer function of the two stages. Harmonic and sub-harmonic peaks persist. c) Response of the array of 250 cells. The AM frequency = 100Hz is well mapped by cells connected to 8ms PRA stellate cells. d) Mapping of mixed signals with AM frequency = 100Hz + 120Hz. A 20% frequency difference is well resolved. The isolated response c) is superimposed, showing the independence of the two peaks.

5.2.2 Inferior Colliculus Level

Stellate cells in the cochlear nucleus code for both the stimulus mean intensity level and temporal fluctuations in the low frequency domain. We could analyze the temporal information with or without using an *a priori* knowledge of the stimulus frequency. The representation is *relative or absolute*. Examples of absolute methods are: (1) extraction by thresholding and (2) temporal derivation. We apply an adaptive threshold for modeling the absolute AM frequency rate-place coding at the inferior colliculus (IC) level (Figure 5.2). The residue of stellate cell activity over the threshold is integrated over a long time period, and the mean rate is computed, giving a value in spikes/second: the mean over threshold (MOT). We suppose that the temporal information, expressed by the synchrony index at the cochlear nucleus level, is completely converted into an absolute value, namely a mean discharge rate, at the inferior colliculus level. Rate-place coding of the envelope frequency is a common property with the autocorrelation.

In order to obtain great simplifications without reducing the descriptive and predictive power, we use models without peripheral front end, and we add the minimal operators necessary for a given function. The input AM signals of our simulations are just represented by their envelopes because the white noise, neural or present in the signal, is modeled by constant probabilistic values. We do not specify the degree of convergence necessary for extracting the fine temporal information of the signal, which could be evaluated separately. Finally, an array of such couple of units (stellate and IC cells) could constitute a bank of independent neural filters just by varying the PRA of stellate cells. This allows us to perform the segregation of two overlapped AM sources without using autocorrelation (Figure 5.2).

Consequently, a periodicity analysis could be roughly achieved by a bank of band-pass filters distributed among the modulation frequency domain. Moreover, a major simplification of this filtering method is obtained if we directly apply a FFT after peripheral filtering and rectification of the signal, for evaluating the Fourier Transform magnitude of the envelope. This produces a representation of AM frequency compatible with the mapping observed by Langner and Schreiner (1988). We also successfully developed a functional model of vowels segregation based on this principle (Berthommier, 1992, Berthommier & Meyer, 1995), giving recognition scores similar to Assmann and Summerfield's (1990). But we think that the information processing occurring in the auditory system could be modeled more precisely, in order to understand the temporal mechanisms and the dynamicity of neural integration better.

5.3 PSYCHOACOUSTICAL EXPERIMENTS

We achieved a set of psychoacoustical experiments related with these models of the tonic channel showing that it could carry out a periodicity analysis based on neural resonance in response to a driving amplitude modulation. Comparison between modeled and psychophysical data are achieved for:

(1) modulation-detection absolute *threshold*

(2) *masking* between two AM white noises

(3) sensitivity to *dephasing*

These experiments are complementary. The first one displays the relationships between probabilities of correct response of the subject and the main properties of auditory neurons: threshold, synchronization and stochasticity. Then, masking curves are established for cochlear nucleus neuron models and for subjects, showing a clear similarity when we look at the neurons related, directly or harmonically, to the signal frequency. Third, we show a sensitivity to dephasing which is a feature of a system having channels carrying several interacting components. On the contrary, this is lost by a fully selective system, or by mechanisms removing the phase. These three points demonstrate a mechanism establishing dependencies specific to harmonically related components responses. Such dependency is only observed for the sensitive neural population whereas responses to distant and nonrelated components remain independent.

5.3.1 AM Sensitivity

AM detectability was estimated by modeling a single neuron located in the central nucleus of the inferior colliculus. A single "neurometric" function for AM detection at one modulation frequency was generated using a two-interval, two-alternative forced choice paradigm. On each trial of the experiments, AM was taken to be correctly detected by a stochastic model if the number of spikes in response to the modulated signal exceeded the number of spikes in an otherwise identical interval that contained an unmodulated signal. *Psychometric* functions for four subjects and one modulation frequency were also measured under the same stimulus conditions. Comparison of the simulated neurometric and psychometric functions suggested that there was sufficient information in the rate response of a well-tuned IC neuron to support behavioral detection performance (Lorenzi et al., 1994).

5.3.2 Masking

Psychoacoustical masking modulation curves are established for four subjects in the low modulation-frequency domain (25Hz~800Hz). We use an adaptive two intervals, two-alternative forced choice procedure to measure the threshold of the signal modulation which is an AM white noise at a given frequency (100Hz). The masker consists of an added AM white noise with variable frequency. Each trial of the experiment has an interval with the masker alone, and another interval with masker + signal. These curves always have a main gap related to the signal frequency and a subharmonic gap whereas the octave gap is attenuated (Figure 5.3d).

The complete model of these masking patterns is not done at the systemic level, but a first interpretation is that AM maskers near signal frequency are more effective because they recruit the neural population having the same BMF as the signal, according to the band-pass filtering properties of cochlear nucleus neurons (Frisina et al., 1990). At the neuronal level, we initially determined the strength of the HH model synchronous response to white noises modulated at two different frequencies around the BMF and added with various intensity ratios. We show the enhancement of AM separation of a well-tuned AM

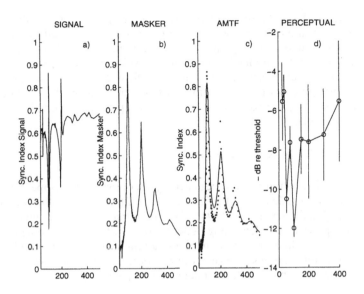

FIG. 5.3. Masking patterns at the cochlear nucleus level and perceptual pattern. The stimulus is the sum of two AM components. The first one (signal) is fixed at 100Hz and the second one (masker) varies among the low frequency domain (horizontal axis of the 4 curves). Modulation depth of signal = 0.25, masker = 0.5. a) The synchrony index of a (modeled) cell tuned to 100Hz shows additive and subtractive peaks (i.e. gaps) related to the AM frequency of the signal, but also to the octave and sub-octave ones. We must notice that the dephasing between components, here set to zero, significantly modifies the shape of this curve. b) Masker synchrony index of the same cell for the same stimulus. c) AM transfer function of the cell for the masker alone superimposed with the previous curve, showing a small difference. d) Mean perceptual modulation masking pattern obtained with 4 subjects. Between-subjects standard deviation is plotted with mean performance. AM signal frequency is 100Hz, modulation depth of masker = 0.5. In order to obtain a gap the phase is PI/2 only for the 100Hz masker. Values are given in dB relative to signal threshold. Nonmasked threshold = -19.53dB, s.d. = 0.43dB. The suboctave gap is always observed.

frequency added to another, relative to the auditory-nerve fibers inputs (Lorenzi & Berthommier, 1994). This corresponds to the BMF band-pass modulation gain of the neurons observed for high intensity levels.

More interesting is the presence of a second subharmonic gap in the perceptual masking pattern, showing that harmonically related signal and masker could strongly interfere in the physiological representation, probably already at the *neural* level. The signal could activate both BMF and subharmonic BMF neurons (Figure 5.1 and 5.2), but these octave related populations are recruited at the same time by the masker. This explains the presence of multiple peaks and gaps in modeled masking patterns (Figure 5.3).

We conclude that: (1) the signal could be masked at the unit level by neighboring or harmonic frequencies (2) the response of this cell for the masker does not depend on the signal. Hence, the interaction between the two components is not symmetric at the unit level. The comparison between perceptual and neural responses becomes conclusive when we use the relative representation given by the synchrony index (Figure 5.3a and 5.3d), but additive contributions of signal and masker are not well separable at the unit level in the absolute MOT map.

5.3.3 Sensitivity to Dephasing

The stimulus is a white noise modulated by a composite envelope built with a low frequency f0 added with the first harmonic 2f0.[i] The paradigm is inspired by Strickland and Viemeister (1994). We show that there exist perceptual AM detection gaps when the dephasing between f0 and 2f0 gives a global composite envelope less efficient for driving neurons having a f0 BMF. A complementary experiment shows the absence of these perceptual gaps when we use f0+3f0. This is also verified by modeling stellate cells responses.

5.4 IMPLICATIONS FOR MODELING

In order to group individual components of a mixture of complex sounds and to well separate concurrent sources on the basis of periodicity analysis, three methods are now available:

(1) autocorrelation

(2) envelope FFT coupled with sieve

(3) bank of it multi-harmonic filters

Autocorrelation has proved its efficiency at the functional point of view, and it can also explain a variety of psychoacoustical phenomena. We cannot discard it with a simple argumentation because it could be complementary with the envelope processing we describe here. But we think that it is more related to pitch identification than to envelope shape sensitivity. Harmonic grouping property is shared by both methods. Let us explain this in more detail.

5.4.1 Autocorrelogram and Physiology

Extensive autocorrelation-based representations of the pitch could be lodged at the cortical level, but no direct observation is done. It is currently admitted that the fine temporal information is impaired by the transfer to upper levels. This

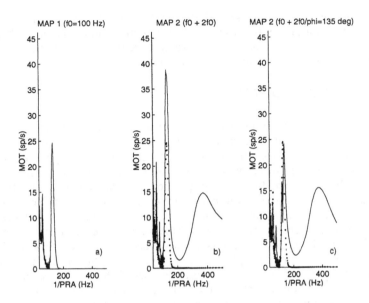

FIG. 5.4. Response to f0+2f0 at the IC level. The map a) is the same as Fig. 5.2c response to f0 = 100Hz. b) This first response map is superimposed with the mapping of 100Hz + 200Hz without dephasing. We show that the response to the fundamental frequency f0 is enhanced, suggesting a physiological grouping mechanism of harmonic components. The main peak, related to f0, is a superposition of responses to both components c) but this response is sensitive to the relative dephasing between them. The enhancement of the main peak completely disappears at 135 degrees.

implies a low level processing which is not necessarily a rate-place recoding. An other appealing strategy could be to reinforce these temporal features at the brain-stem level. Onset cells are candidates for such process by precisely coding the intervals between periods because these cells emit impulses well phase-locked with the envelope (Meyer & Ainsworth, 1993).

5.4.2 Phase Removing

It is interesting to see that FFT-based models remove the dephasing like an autocorrelogram. The autocorrelation is just the inverse Fourier transform of the squared Fourier transform magnitude. Consequently, both autocorrelogram and FFT are not sensitive to the fine shape variations of a *composite envelope* induced by the dephasing between two octave related components. On the contrary, the physiological implementation we propose well responds to these variations (Figure 5.4). The autocorrelation alone cannot explain the psychoacoustical gaps of sensitivity produced by this dephasing, whereas it is compatible with the responses of the bank of multiharmonic filters. The discussion is not closed because a poorly selective model having a threshold could also explain a sensitivity to such a dephasing, and the onset cell population is also sensitive to the same variations.

5.4.3 Pooling Strategies

Neural filters are not simple resonators but multi-bandpass filters sensitive to both a low fundamental frequency and at least the first harmonic: cells are directly sensitive to *harmonicity*. This suggests a *temporal* grouping process alternative to the sieve when components are harmonically related. Autocorrelogram also directly groups the harmonic components, but it is necessary to detect the main peaks in a summarized representation for identifying the fundamental frequencies of mixed signals. A large scale integration is performed somewhere over the tonotopic representation in order to identify the pitch. Each spectrum is directly place-coded facing these peaks, and the use of the sieve for a supplementary sampling is optional. On the contrary, the FFT evaluation of envelope spectrum performed channel by channel must be coupled with the sieve in order to recover the fundamental frequencies because the energy is dispersed among harmonics. It is applied on the summary envelope spectrogram. Each spectrum is recovered with an explicit pooling process based on the precise knowledge of the fundamental frequency (Berthommier & Meyer, 1995). The computational cost is lower for FFT than for autocorrelation. The multiharmonic filtering scheme we propose seems interesting because:

(1) Harmonic sensitivity is explained by the temporal coherence

(2) Pooling harmonics does not need a previous identification of the fundamental frequency

(3) Resonances on the fundamental frequency are obtained without heavily blurring the spectral information

(4) Dephasing between harmonics is not removed, preserving the fine details of the integrative processes

(5) It is well related with the FFT + sieve method

This does not exclude a complementarity between neural filtering and autocorrelation depending on the task and the level of processing.

5.5 ACKNOWLEDGMENTS

This work was supported by CNRS, "Groupe Perception et reconnaissance de formes du Pôle Rhône-Alpes de Sciences Cognitives" and is a part of the SPHERE HCM contract. Thanks to J. L. Schwartz for helpful discussions, to the paper referee for useful criticism and to C. Micheyl for the software assistance in psychoacoustical experiments.

REFERENCES

Assmann, P. F. & Summerfield, Q. (1990). Modeling the perception of concurrent vowels: Vowels with different fundamental frequencies, *JASA*, 88: pp. 680-697.

Berthommier, F., Schwartz, J. L. & Escudier, P. (1989). Auditory processing in a post-cochlear neural network: Vowel spectrum processing based on spike synchrony. In, *Proceedings of Eurospeech 89*, J. P. Tubach and J. J. Mariani (Eds.), pp. 247-250, Paris.

Berthommier, F. (1992). Intégration neuronale dans le système auditif, Thèsc GBM-USMG, Grenoble.

Berthommier F. & Meyer, G. (1995) Source separation by a functional model of amplitude demodulation. In, *Proceedings of Eurospeech 95*, J. M. Pardo et al. (Eds.), pp. 135-138, Madrid, Spain.

Bregman, A. S. (1990). *Auditory Scene Analysis*. Cambridge, MA: MIT Press.

Cooke, M., Brown, G. J., Crawford, M. & Green, P. (1993). Computational auditory scene analysis: listening to several things at once. In *Endeavour*, 17(4): pp. 186-190, New Series, Pergamon Press.

Frisina, R. D., Smith, R. L. & Chamberlain, S. C. (1990). Encoding of amplitude modulation in the gerbil cochlear nucleus: I. A hierarchy of enhancement, *Hear. Res.*, 44: pp. 99-122.

Hewitt, M. J., Meddis, R. & Shackleton, T. M. (1992). A computer-model of cochlear nucleus stellate cell: responses to amplitude modulated and pure tone stimuli, *JASA*, 91: pp. 2096-2109.

Langner G. & Schreiner, C. E. (1988). Periodicity coding in the inferior colliculus of the cat. I: Neuronal mechanisms, *J. Neurophysiol.*, 60: pp. 1799-1822.

Lorenzi, C., Micheyl, C. & Berthommier, F. (1994). Sensitivity to amplitude modulation: neurobiological modelling and psychoacoustical experiments.

In, *Proceedings of Conf. on Neural Modelling*, AIDRI, M. Brissaud (Ed.), pp. 111-115, Lyon, France.

Lorenzi C. & Berthommier, F. (1994). A computational model for amplitude modulation extraction and analysis of simultaneous amplitude modulated signals, *Journal de Physique*, Suppl. au J. de Phys. III, 4(C5): pp. 379-382.

Meyer G. & Ainsworth, W. A. (1993). Vowel pitch period extraction by models of neurones in the Mammalian brain-stem. In, *Proceedings of Eurospeech 93*, 3: pp. 2029-2033.

Schwartz, J. L., Beautemps, D. , Arrouas, Y. & Escudier, P. (1992). Auditory analysis of speech gestures. In, *The Auditory Processing of Speech*, M. E. H. Schouten (Ed.), pp. 239-252, Mouton de Gruyter, Berlin.

Strickland, E. A. & Viemeister, N. F. (1994). What aspects of the envelope are relevant for detection of amplitude modulation?, *JASA*, 95: 2964.

NOTE

[i] This experiment was done in collaboration with L. Demany; lab. de Psychoacoustique, Bordeaux.

6

Stream Segregation Based on Oscillatory Correlation

DeLiang Wang
The Ohio State University

Auditory segmentation is critical for complex auditory pattern processing. We present a neural network framework for auditory pattern segmentation. The network consists of laterally coupled two-dimensional neural oscillators with a global inhibitor. One dimension represents time and another one represents frequency. We show that this architecture can in real time group auditory features into a segment by phase synchrony and segregate different segments by desynchronization. The network demonstrates a set of psychological phenomena regarding primitive auditory stream segregation. The neuroplausibility and possible extensions of the model are discussed.

6.1 INTRODUCTION

At any time a listener is being exposed to acoustic energy from many simultaneous auditory events. In order to recognize and understand such a dynamic environment, the listener must first disentangle the acoustic wave and capture each event. This process of auditory segmentation is referred to as auditory stream segregation or auditory scene analysis (Bregman, 1990), a critical part of auditory perception.

Auditory segmentation was first reported by Miller and Heise (1950) who noted that listeners split the signal with two sine wave tones into two segments. Bregman and his collaborators carried out a series of studies on this subject. In one of the early studies, subjects were asked to report the temporal order of six tones in the sequence. Three of them were in a high frequency range, and the other three in a low frequency range. This situation is simplified into Fig. 6.1.

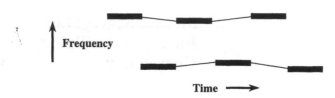

FIG. 6.1. Six alternating pure tones as displayed in a spectrogram. When stream segregation occurs, the high-frequency tones form one stream and the low-frequency tones form another stream (indicated by thin lines).

The results showed that at high rates of presentation, subjects perceived two separate sequences corresponding to the high and low frequency tones respectively, and they were able to report only the temporal order of the tones within each sequence, but not across the two sequences. This basic phenomenon of stream segregation was repeatedly verified in different contexts (Bregman, 1990; Jones, 1976). In general, if auditory patterns are displayed on a spectrogram, the results are consistent with Gestalt laws of grouping that were expressed in the visual domain.

von der Malsburg and Schneider (1986) proposed perhaps the only neural network model that addressed the problem of auditory segmentation. They described the idea of using neural oscillations for expressing segmentation, whereby a set of features forms the same segment if their corresponding oscillators oscillate in synchrony, and oscillator groups representing different segments desynchronize from each other. Using a fully connected oscillator network, they demonstrated segmentation based on onset synchrony, that is, oscillators simultaneously triggered (a segment) synchronize with each other. Generally, a fully connected network indiscriminately connects all the oscillators which are activated simultaneously by different objects, because the network is dimensionless and loses critical geometrical information. Because of this, their model could not extend very far. For example, the model cannot demonstrate stream segregation that requires an account of frequency proximity. Computer algorithms have been developed to separate two speakers or two events on the basis of different fundamental frequencies or other cues (Parsons, 1976; Weintraub, 1986; Beauvois & Meddis, 1991; Brown & Cooke, 1994). The success of these models is quite limited, and it is not clear how the models could be extended to handle real time sound separation based on a variety of grouping principles.

In the following, we present a neural model for auditory stream segregation. Similar to the model of von der Malsburg and Schneider (1986), our model is based on the idea of oscillatory correlation, whereby phases of neural oscillators encode the binding of sensory components. However, both the single oscillator model and the neural architecture are fundamentally different from those used by von der Malsburg and Schneider. Simulations show that the model is capable of replicating the basic phenomenon of stream segregation and explaining a set of psychological observations concerning primitive auditory scene analysis. The framework proposed here promises to explain a variety of experimental observations and to provide an effective computational approach to auditory segmentation (see also Wang, 1994). For a more extended version see Wang (1996).

6.2 NEURAL ARCHITECTURE

We introduce a network of neural oscillators to model primitive auditory segmentation, called the segmentation network. The building block of the segmentation network, a single oscillator i, is defined as a feedback loop between an excitatory unit xi and an inhibitory unit yi (cf. Wang & Terman, 1995; Terman & Wang, 1995)

$$\frac{dx_i}{dt} = 3x_i - x_i^3 + 2 - y_i + I_i + S_i + \rho \qquad \text{EQ. 6.1a}$$

$$\frac{dy_i}{dt} - \varepsilon(\gamma(1 \mid Tanh(x_i / \beta)) - y_i) \qquad \text{EQ. 6.1b}$$

where I_i represents external stimulation, S_i represents overall coupling from other oscillators, and ρ denotes the amplitude of a Gaussian noise term. The parameter ε is chosen to be small. In this case (1), without any coupling or noise, corresponds to a standard relaxation oscillator. The x-nullcline of (1) is a cubic curve, while the y-nullcline is a sigmoid function with the parameters β and γ. For $I > 0$, (1) gives rise to a stable periodic orbit for all values of ε sufficiently small. The periodic solution quickly alternates between a phase of relatively high values of x, called the *active phase* of the oscillator, and a phase of relatively low values of x, called the *silent phase* of the oscillator. For $I < 0$, (1) produces a stable fixed point.

Time plays a critical role in stream segregation. We treat time in this model as a separate dimension. To simplify the discussions, we consider only time and frequency: rows represent time and columns represent frequency, as shown in Fig. 6.2. Each oscillator in the matrix is laterally connected with its neighboring oscillators. The global inhibitor receives excitation from each oscillator, and inhibits in turn each oscillator. The network has an input end which consists of units representing distinct frequencies, called *input channels*. Each input channel connects to a corresponding oscillator row representing the same frequency, called a *frequency channel*, by delay lines with different delays. These delay lines are arranged so that delays increase systematically from left to

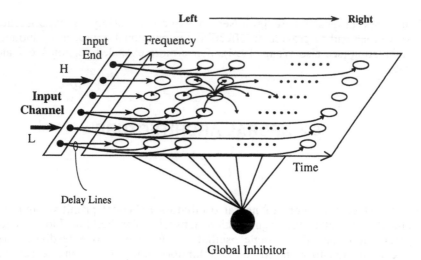

FIG. 6.2. Diagram of the segmentation network. The connections from a typical oscillator in the network are shown in the figure, and those from other oscillators are omitted for clarity. As in the following figures, symbol H indicates an auditory input with a high frequency and L indicates an auditory input with a low frequency.

right. Thus, each oscillator in the matrix is activated by an input of a specific frequency at a specific time relative to the present time.

Inspired by the idea of dynamic links (von der Malsburg & Schneider, 1986), we recently introduced a mechanism called *dynamic normalization* (Wang, 1993; Wang, 1995). In this scheme, there is a pair of connection weights from oscillator j to i, one permanent T_{ij}, and another dynamic J_{ij}. Permanent links reflect the hardwired structure of a network, while dynamic links quickly change their strengths from time to time, depending on the current state of the network. More specifically, dynamic normalization ensures that each oscillator has equal overall dynamic connection weights (J_{ij}) from all its input oscillators. Dynamic normalization is assumed in this model.

The permanent connectivity pattern between the oscillators in the segmentation network, except for the self connection, is assumed to take on a two dimensional Gaussian distribution. Let the two dimensional indices of oscillator i be (t_i, f_i), representing the time and frequency coordinates of the oscillator respectively. Oscillator i connects to oscillator j with strength

$$T_{ij} = Exp[-(\frac{(t_j - t_i)^2}{\sigma_t^2} + \frac{(f_j - f_i)^2}{\sigma_f^2})]$$

EQ. 6.2

where the parameters σ_t and σ_f determine the widths of the Gaussian distribution along the time axis and the frequency axis, respectively. The self connectivity T_{ii}'s is set to 0. Once T_{ij}'s are defined, J_{ij}'s are updated according to dynamic normalization.

The coupling term S_i to oscillator i is given by

$$S_i = \sum_j J_{ij} S_\infty(x_k, \theta_x) - W_1 S_\infty(z_1, \theta_1) - W_2 S_\infty(z_2, \theta_2) \qquad \text{EQ. 6.3}$$

$$\frac{dz_1}{dt} = \phi(\sigma_\infty - z_1) \qquad \text{EQ. 6.4a}$$

where $S_\infty(x, \theta) = 1/(1 + Exp[-\kappa(x - \theta)])$. The first term of the right-hand side of (3) describes the lateral connections to i, and the second term and the third term describe inhibition from the global inhibitor. W_1 and W_2 are the weights of inhibition from the global inhibitor which we denote by a pair of the units z_1 and z_2, defined as

$$\frac{dz_2}{dt} = \phi(\sigma_0 / N_t - z_2) \qquad \text{EQ. 6.4b}$$

where $\sigma_\infty = 0$ if $x_i < \theta_z$ for every oscillator i and $\sigma_\infty = 1$ if $x_i \geq \theta_z$ for at least one oscillator i, and σ_0 equals the number of oscillators whose x activities are greater than or equal to θ_z. If the x activity of every oscillator is below θ_z, both z_1 and z_2 approach 0, and the oscillators on the network receive no inhibition. On the other hand, if the x activity of at least one oscillator is above the threshold, the global inhibitor will receive input. In this case, z_1 approaches 1, and z_2 equals the number of supra threshold oscillators divided by N_t, which is the total number of the oscillators in a row of the segmentation network (Fig. 6.2). θ_2 is set to $2/N_t$. Thus, every oscillator on the network will sense inhibition. Finally, ϕ determines the rate at which the inhibitor reacts to such stimulation. θ_x, θ_z, and θ_1 are all thresholds.

The system of (1)-(4) in a slightly simplified version has recently been analyzed in general contexts by Terman and Wang (1995) (for a much abbreviated version see Wang & Terman, 1995). This network exhibits a mechanism of *selective gating*, whereby an oscillator jumping up to the active

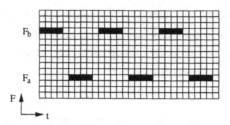

FIG. 6.3. A stimulus pattern as mapped to the segmentation network at a specific time. This condition corresponds to fast presentation and large frequency separation.

phase rapidly recruits the oscillators stimulated by the same pattern, while preventing other oscillators from jumping up. With the selective gating mechanism, the network rapidly achieves both synchronization within each pattern and desynchronization between different patterns. Contrary to fully connected ones, locally coupled oscillator networks of this kind preserve geometry in input patterns. The mechanism of selective mechanism provides the computational foundation for our study of auditory stream segregation.

6.3 SIMULATION RESULTS

The segmentation network (Fig. 6.2) has been simulated with respect to auditory segmentation. To reduce numerical computations involved in integrating a large number of the differential equations, a computer algorithm has been extracted, which veridically follows every major step in the numerical simulation of the equations. In order to relate to real time, we assume that the basic delay interval, that is, the difference of delay between two neighboring isofrequency oscillators is 40 ms. When an oscillator is triggered, a random phase is generated for it, and the phases of all activated oscillators are randomized after every delay step.

6.3.1 Stream Segregation

To simulate the basic phenomenon of stream segregation, a sequence of six alternating tones *HLHLHL* is used as input. All L tones were assumed to trigger the same frequency channel (F_1), and so were all H tones (F_2). The distance between F_1 and F_2 was set to eight rows, corresponding to large frequency separation. The rate of presentation was 4 delays per tone, corresponding to fast presentation. The sequence was repeatedly presented to the network, as in the psychological experiments. A network of 15x30 oscillators was simulated, as shown in Fig. 6.3. Fig. 6.4 displays the complete response of the F_1 and F_2 channels. Because other frequency channels were not stimulated, the oscillators in those channels were always silent, hence omitted in the display. Each trace displays the activity of the excitatory unit of one oscillator. The top 30 traces represent the 30 isofrequency oscillators with progressively increasing response delays in the F_2 channel (*H* tone). Similarly, the bottom 30 traces represent the 30 oscillators in the F_1 channel (*L* tone). A total of 40 delay steps were simulated, while the complete sequence of the six tones corresponds to 30 delay steps (see Fig. 6.3). The vertical lines were included to help compare the phases of different oscillators. As can be seen from the figure, except for a beginning period, all active oscillators of the F_2 channel rapidly reached synchronization, and so did the oscillators of the F_1 channel. Furthermore, the oscillators of one channel desynchronized with those of the other channel. Taken together, all *H* tones are grouped into one segment, and all *L* tones are grouped into another segment. Relating to psychological experiments, stream segregation occurred for fast presentation with large frequency separation, and two streams were segmented apart in *real time*.

Fig. 6.5 shows the combined x values (see Equation 1) of all oscillators of each of the frequency channels for one typical delay interval after the presentation of a full sequence of six tones was completed. The top and the middle panels show the activities of the F_2 and F_1 channels respectively, and the bottom panel shows the activity of the global inhibitor during the same time period. For each frequency channel, only the activated oscillators are included in the display since inactivated oscillators are always silent. As is clear in the figure, the quality of synchronization within the same frequency channel improved after the first cycle of oscillations. The frequency of the global inhibitor is double that of an activated oscillator in the segmentation network, since the inhibitor is activated by both segments. Though rather rigid due to algorithmic implementation, synchronization within the same frequency channels and desynchronization across the two channels are clearly captured in the format of illustration of Fig. 6.5, which will be used for all the following simulations.

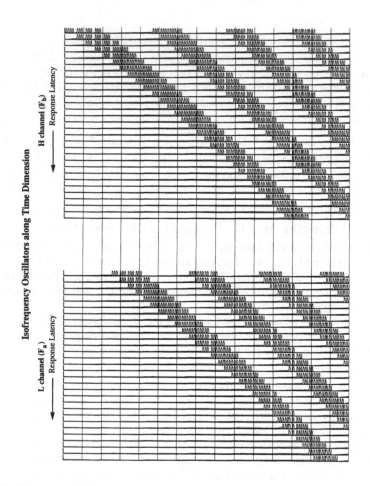

FIG. 6.4. Response of the two corresponding frequency channels to fast presentation of alternating H and L tones with large frequency separation. Each activity trace represents the normalized value of the excitatory unit of an oscillator. The parameter values used: k = 50, qx = -0.5, qz = 0.1, q1 = 0.5, W1 = 0.5, W2 = 1.0, st = 8.0, sf = 5.0, Ii = 0.2 if oscillator i is externally stimulated, and Ii = -0.02 otherwise. Dynamic links were normalized to 6.0. The algorithm took 1,500 steps.

We also observed that when the frequency separation between the two frequency channels was reduced to 4 rows (cf. Fig. 6.3), no segregation occurred. In this case, all of the stimulated oscillators of both channels were synchronized, and all of the *L* and *H* tones were grouped into the same stream. Furthermore, with large frequency separation but when the rate of tone representation was reduced by half, each tone became a single segment and no grouping occurred at all. The same phenomenon of stream segregation occurs when tones do not have constant frequencies shown in Fig. 6.3, but are FM

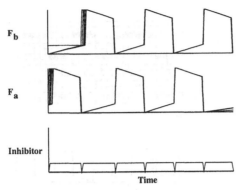

FIG. 6.5. The combined activities of all the activated oscillators during one delay interval in the high- (F2) or the low-frequency (F1) channels, respectively. The ordinates indicate the normalized x values of the oscillators and the value of the global inhibitor.

tones. Detailed illustrations are omitted for space. Our simulation results demonstrate that frequency proximity plays the dominant role in stream segregation, and it is not important whether or not the tones have constant frequencies.

6.3.2 Sequential Capturing

One of the well-known phenomena in auditory scene analysis is so-called sequential capturing. It was first reported by Bregman and Pinker (1978) who tested a repeating sequence formed by a pure tone T_3 and a complex tone composed of two pure tone components T_1 and T_2 (Fig. 6.6A). By varying the frequencies of T_2 and T_3 and the onset time differences between T_1 and T_2, they show that T_3 may capture T_2 from the complex T_1/T_2 to form a new segment T_3/T_2. Sequential capturing is promoted by decreasing frequency distance between T_2 and T_3, increasing frequency distance between T_1 and T_2, or increasing onset time differences between T_1 and T_2.

We have simulated the phenomenon of sequential capturing. As in the experiments, a repeating sequence of three simulated pure tones were used, as shown in Fig. 6.6A. Relating to the experiments, F_a tones correspond to T_1 tones, F_c to T_2, and F_d to T_3. The tones with frequencies F_a and F_c constitute a complex tone, with substantial overlapping in time. We first tested the case that the distance between T_2 and T_3 is closer than that between T_1 and T_2. A network of 15x27 oscillators was simulated so that it could represent a sequence of three repetitions of the stimuli. Fig. 6.6B provides the simulation results within a typical delay step after the whole sequence was presented. As is clear in the figure, the activated oscillators in the F_c and F_d frequency channels were synchronized, and their oscillations were desynchronized from those in the F_a channel. In this case, the T_3 tones captured the T_2 tones from the T_1/T_2 complex to form a new segment T_3/T_2. Next we tested the case where the frequency

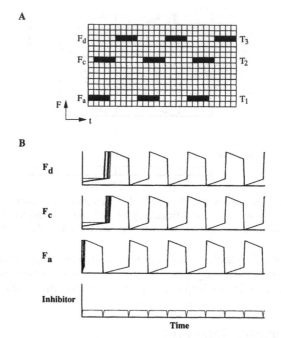

FIG. 6.6.   Sequential capturing: Case 1. A: The stimulus pattern as mapped to the segmentation network at a specific time.   Overlapping F_a and F_c tones compose the complex tone and F_d tones are the captor.   The separation between F_a and F_c is 7 rows, and between F_c and F_d is 4 rows.   B: The combined activities of all of the activated oscillators in each of three frequency channels, plus the activity of the global inhibitor. Only shown is one delay interval after the full sequence of six tones was presented.   The parameter values are the same as in Fig. 6.4 except that W1 = 0.3 and W2 = 1.2.   The algorithm took 1,600 steps.

separation between T_2 and T_3 is larger than that between T_1 and T_2, as shown in Fig. 6.7A.   The simulation results are shown in Fig. 6.7B, which clearly demonstrate that the complex tone of T_1/ T_2 is kept together and it is separated from the captor T_3. That is, no capturing was exhibited in the simulation.

The simulation results in Figs 6.6 and 6.7 resemble those of the corresponding psychological experiments (Bregman & Pinker, 1978; Bregman, 1990).   Different from stream segregation, these simulations of sequential capturing used tones with time overlaps.   As shown in the results, unlike other models of auditory segmentation (e.g. Beauvois & Meddis, 1991), our model handles simultaneous tones equally well as sequential tones.

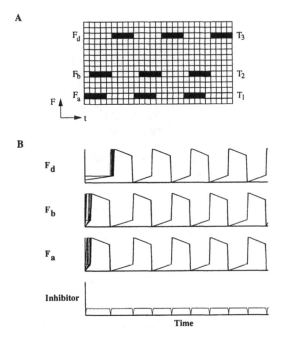

FIG. 6.7. Sequential capturing: Case 2. The only difference from Fig. 6.6 is that T_2 tones in this case have frequency F_b instead of F_c.

6.3.3 Competition Among Alternative Organizations

Whether a set of tones is grouped into the same segment depends not only on the arrangement within the set but also on the context of which the set is part. Grouping takes place as though different segments competed for the belonging of a specific tone. To test whether our model exhibits appropriate competition among rival organizations, we have simulated the experiment of Bregman (1978). In the experiment, two tones T_1 and T_2 with a fixed frequency separation formed the first pair and were presented successively. At the same time, another two tones T_3 and T_4, also with fixed frequency separation, formed the second pair and were also presented successively. The presentation of the two pairs was interleaved in time. When the frequency separation between the pairs was large, each pair formed its own stream (segment). But when the two pairs were brought into the same frequency region so that the frequency distance within each pair was greater than the distances between T_1 and T_3 and between T_2 and T_4 across the pairs, then T_1 of the first pair formed a segment with T_3 of the second pair and T_2 formed a segment with T_4.

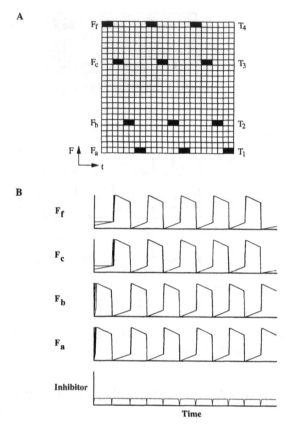

FIG. 6.8. Competition among different organizations: Case 1. A: The stimulus pattern as mapped to the segmentation network at a specific time. The stimulus is a sequence of 12 tones, triggering four frequency channels: F_a, F_b, F_c, or F_f. B: The combined activities of all of the activated oscillators in each of four frequency channels, plus the activity of the global inhibitor. Only shown is one delay interval after the full sequence of six tones was presented. The parameter values are the same as in Fig. 6.4 except that W1 = 0.1 and W2 = 1.8. The algorithm took 1,600 steps.

To simulate this experiment, a network of 24x24 oscillators was used. Following the experiment, each tone was assumed to be brief, occupying only two oscillators. We first tested the case with large frequency separation between the two pairs. The stimulus condition is described in Fig. 6.8A, where the first pair of T_1/T_2 stimulates the frequency channels of F_a and F_b respectively, and the second pair of T_3/T_4 stimulates the frequency channels of F_c and F_f respectively. The frequency distance between F_a and F_b is 5 rows, between F_c and F_f is 7 rows, and between F_b and F_c is 11 rows. Fig. 6.8B displays the simulation results. The figure shows that the sequence of T_1/T_2 tones (the first pair) was grouped into

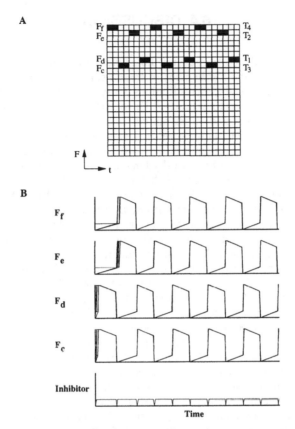

FIG. 6.9. Competition among different organizations: Case 2. The only difference from Fig. 6.8 is that in this simulation T_1 and T_2 tones trigger F_d and F_e channels, respectively.

the same segment by phase synchronization, and the sequence of T_3/T_4 tones (the second pair) was grouped into a different segment. We later moved the first pair up to the same frequency region as the second pair, while keeping the frequency distance within each pair unchanged. In this case, the first pair of T_1/T_2 stimulated the frequency channels of F_d and F_e respectively, as shown in Fig. 6.9A. The simulation results are given in Fig. 6.9B. In this case, T_1 tones were grouped with T_3 tones, instead of T_2 tones. T_2 tones were grouped with T_4 tones. The two pairs were broken down into different segments. The simulation results well replicate the experimental findings of Bregman (1978), and capture the essential properties of competition among alternative organizations.

These simulations demonstrate that the behavior of the segmentation network when performing a variety of primitive stream segregation tasks resembles relevant psychological observations. We conclude that tones can be grouped together based on their proximity in frequency, their closeness in time, and the auditory context within which tones occur.

6.4 DISCUSSION

There is ample evidence suggesting the existence of neural oscillations in the brain. The range of the frequencies of these oscillations is between 20 and 80 Hz (often referred to as 40 Hz oscillations, Gray et al., 1989). Auditory evoked potentials were observed to exhibit 40 Hz oscillations (Galambos et al., 1981). Ribary et al. (1991) reported 40-Hz activity in localized brain regions both at the cortical level and at the thalamic level in the auditory system. These oscillations are synchronized over considerable cortical areas, and the synchronized oscillations can be elicited by both rhythmic and transient sound stimuli. Llinás and Ribary (1993) further described that 40 Hz oscillations can be triggered by frequency modulated tones. An important element of the architecture of the model is the use of shift circuits with delay lines. Time delays of neuronal responses have been found at various levels of the visual pathway, and it appears that delays become longer in higher auditory structures. In the cat auditory cortex, electrophysiological recordings identify up to 1.6 second delays in response to identical tones separated by certain periods or a sequence of different tones (McKenna et al., 1989). Our model gives rise to a novel neurocomputational theory, called the oscillatory correlation theory, that could explain how primitive auditory segmentation may be achieved in the brain. The theory will be described in detail elsewhere.

Although Section 3 shows only preliminary simulation results of auditory stream segregation, the model is not limited by the stimuli used. For example, the model can be extended to include another layer for detecting stimulus onset, and grouping based on onset synchrony can be achieved by strengthened connections between onset detectors that are activated more or less at the same time. The permanent connectivity pattern of the Gaussian distribution (cf. (2)) strongly biases towards the grouping of sounds that have continuous frequency transitions, which is consistent with the analysis on speech perception. Of course, the basic architecture of Fig. 6.2 must be extended to incorporate other qualities of auditory stimuli, such as amplitude, rhythm, harmonics, timbre, etc. Grouping based on common amplitude/frequency modulation may be handled in a similar way as onset synchrony. Grouping of multiple frequency partials that form harmonics of a fundamental frequency is based on two cues. The first is onset/offset synchrony among these partials. The second is spectral relations among the partials. There are many models for pitch perception on the basis of harmonic relations. Despite these considerations, the basic principles of this model may remain the same.

6.5 ACKNOWLEDGMENTS

The preparation of this chapter was supported in part by an ONR grant N00014-93-1-0335 and NSF grants IRI-9211419 and CDA-9413962.

REFERENCES

Beauvois, M. W. & Meddis, R. (1991). A computer model of auditory stream segregation. *Quarterly Journal of Experimental Psychology*, 43 A, 517-541.

Bregman, A. S. (1978). Auditory streaming: Competition among alternative organizations. *Perception and Psychophysics*, 23, 391-398.

Bregman, A. S. (1990). *Auditory Scene Analysis*. Cambridge, MA: MIT Press.

Bregman, A. S. & Pinker, S. (1978). Auditory streaming and the building of timbre. *Canadian Journal of Psychology*, 32, 19-31.

Brown, G. & Cooke, M. (1994). Computational auditory scene analysis. *Computer Speech and Language*, 8, 297-336

Galambos, R , Makeig, S & Talmachoff, P. T. (1981). A 40-Hz auditory potential recorded from the human scalp. *Proceedings of the National Academy of Sciences of USA*, 78, 2643-2647.

Gray, C. M., König, P., Engel, A. K., & Singer, W. (1989). Oscillatory responses in cat visual cortex exhibit inter-columnar synchronization which reflects global stimulus properties. *Nature*, 338, 334-337.

Jones, M. R. (1976). Time, our lost dimension: Toward a new theory of perception, attention, and memory. *Psychological Review*, 83, 323-355.

Llinás, R. & Ribary, U. (1993). Coherent 40-Hz oscillation characterizes dream state in humans. *Proceedings of the National Academy of Sciences of USA*, 90, 2078-2082.

McKenna, T. M., Weinberger, N. M., & Diamond, D. M. (1989). Responses of single auditory cortical neurons to tone sequences. *Brain Research*, 481, 142-153.

Parsons, T. W. (1976). Separation of speech from interfering speech by means of harmonic selection. *Journal of Acoustical Society of America*, 60, 911-918.

Ribary, U., Ioannides, A. A., Singh, K. D., Hasson, R., Bolton, J. P. R., Lado, F., Mogilner, A., & Llinás, R. (1991). Magnetic field tomography of coherent thalamocortical 40-Hz oscillations in humans. *Proceedings of the National Academy of Sciences of USA*, 88, 11037-11041.

Terman, D. & Wang, D. L. (1995). Global competition and local cooperation in a network of neural oscillators. *Physica D*, 81, 148-176.

von der Malsburg, C. & Schneider, W. (1986). A neural cocktail-party processor. *Biological Cybernetics*, 54, 29-40.

Wang, D. L. (1993). Modeling global synchrony in the visual cortex by locally coupled neural oscillators. In *Proceedings of the Fifteenth Annual Conference of the Cognitive Science Society* (pp. 1058-1063), Boulder, CO. Hillsdale, NJ: Lawrence Erlbaum Associates.

Wang, D. L. (1994). Auditory stream segregation based on oscillatory correlation. In J. Vlontzos, J.-N. Hwang, & E. Wilson (Eds.), *Proceedings of The IEEE 1994 Workshop on Neural Networks for Signal Processing* (pp. 624-632). IEEE.

Wang, D. L. (1995). Emergent synchrony in locally coupled neural oscillators. *IEEE Transactions on Neural Networks*, 6, 941-948.

Wang, D. L. (1996). Primitive auditory segregation based on oscillatory correlation. *Cognitive Science*, 20, 409-456.

Wang, D. L. & Terman, D. (1995). Locally excitatory globally inhibitory oscillator networks. *IEEE Transactions on Neural Networks*, 6, 283-286.

Weintraub, M. (1986). A computational model for separating two simultaneous talkers. In *Proceedings of IEEE International Conference on Acoustics, Speech, and Signal Processing* (pp. 81-84), Tokyo.

7

Temporal Synchronization in a Neural Oscillator Model of Primitive Auditory Stream Segregation

Guy J. Brown and Martin Cooke
University of Sheffield

A computational model of primitive auditory stream segregation is presented, in which the grouping of peripheral channels is indicated by temporal synchronization in a network of neural oscillators. The model exhibits a number of important emergent properties, including sensitivity to the frequency proximity, temporal proximity and common onset of spectral components. In addition, the model is able to account for the build-up of streaming over time.

7.1 INTRODUCTION

Bregman's influential account of auditory perception (Bregman, 1990) holds that the complex mixture of sounds reaching the ears is subjected to an *auditory scene analysis*. This process occurs in two conceptually distinct stages. In the first stage, sound is decomposed into a collection of sensory elements (features). Subsequently, elements that are likely to have arisen from the same environmental event are grouped to form a perceptual whole, or "stream".

The emphasis that Bregman's account places on the concept of *grouping* raises the following question: how are groups represented and communicated within the auditory system? This key issue, which has largely been overlooked in the literature on auditory scene analysis, is the subject of the computational modelling study described in this paper.

Recent physiological studies suggest that the higher auditory nuclei contain maps of feature detecting cells, in which frequency and some other parameter are represented on orthogonal axes. Parameters that appear to be represented in this manner include amplitude modulation, spectral shape, interaural time difference, and interaural intensity difference (see Moore, 1987 for a review). Generally, a number of sound sources will be simultaneously active in an acoustic environment, leading to superimposed responses in these auditory maps. At a physiological level, then, the question posed above can be phrased as follows: how does the auditory system bind together the responses of neurons that code for features of the same acoustic source?

This issue was called the *binding problem* by Malsburg and Schneider (1986), who also suggested an elegant solution. Their scheme proposes that neurons that are responding to features of the same environmental event are grouped according to the coherence (synchrony) of their temporal responses. This notion is supported by recent physiological studies, which suggest that visual stimuli initiate synchronized neural oscillations (so-called 40 Hz oscillations) across disparate regions of the visual cortex (Gray, König, Engel & Singer, 1989).

Despite the attractiveness of Malsburg and Schneider's scheme, there have been few attempts to exploit temporal synchronization in models of auditory grouping (an exception is Wang's, 1996, recent work). In the remainder of this chapter, we describe a model of auditory stream segregation in which grouping is signaled by temporal synchronization in a simple neural network. It is demonstrated that, despite its simplicity, the model has considerable explanatory power and is able to account qualitatively for a wide range of auditory grouping phenomena.

7.2 THE MODEL

The proposed model consists of four stages. In the first stage, peripheral auditory processing is simulated by a bank of bandpass filters and a model of inner hair cell function. In the second stage of the model, a simulated auditory nerve excitation pattern, derived from the hair cell model, is processed by a map of onset cells. Third, activity in the onset map is used to modify the coupling strengths between neurons in a fully connected neural network. Cells in the network that are strongly connected exhibit a highly synchronized temporal response, whereas cells that are weakly connected show uncorrelated behavior.

The final stage of our scheme is an attentional searchlight (Crick, 1984), although this is not explicitly modeled in the current implementation. The stages of the model are summarized in Figure 7.1 and described in detail below.

7.2.1 Auditory Periphery

The frequency selective properties of the basilar membrane are simulated by a bank of gammatone filters (Patterson, Holdsworth, Nimmo-Smith & Rice, 1988), each of which models the frequency response of a particular point along the cochlear partition. For reasons of computational efficiency, the range of filter center frequencies was restricted to the range of stimulus frequencies used in our simulations. Specifically, 30 gammatone filters were employed, with center frequencies equally distributed between 500 Hz and 2 kHz on the ERB-rate scale of Glasberg and Moore (1990).

The output from the gammatone filterbank is processed by a model of inner hair cell function, which simulates adaptation, saturation and compression phenomena (Cooke, 1993). It should be noted that Cooke's hair cell model only provides a representation of the *envelope* of the auditory nerve response; temporal fine structure is not retained.

Such a simplified hair cell model is adequate for the study described here, because our scheme does not simulate processes that operate on the fine structure of auditory nerve firing patterns (for example, pitch analysis). Additionally, Cooke's model provides a suitable input for the onset cell model described below.

7.2.2 Onset Map

Neurons that respond with a brief burst of activity at the onset of a tonal stimulus are found throughout the higher auditory nuclei. One possible mechanism that accounts for this behavior is an excitatory input to the cell at the start of stimulation, followed by strong inhibition which prevents subsequent activity (Shofner & Young, 1985). This mechanism can be approximated by writing the membrane potential $p(t)$ at time t as a leaky sum of the excitatory and inhibitory inputs to the cell,

$$p(t) = p(t-1)c + Epsp\ r(t) - Ipsp\ r(t-1) \qquad \text{EQ. 7.1}$$

Here, *Epsp* and *Ipsp* are the strengths of the excitatory and inhibitory inputs respectively, $r(t)$ is the hair cell response and c is a constant that determines the rate at which $p(t)$ decays to its resting level. The firing rate of the onset cell, $s(t)$, is determined by the value of the membrane potential when it exceeds a threshold T,

$$s(t) = \begin{cases} p(t) & p(t) > T \\ 0 & \textit{otherwise} \end{cases} \qquad \text{EQ. 7.2}$$

The parameter values employed are *Epsp* = 1.0, *Ipsp* = 6.0, $c = 0.86$ and $T = 0$. For a more detailed account of the onset cell model, see Brown and Cooke (1994).

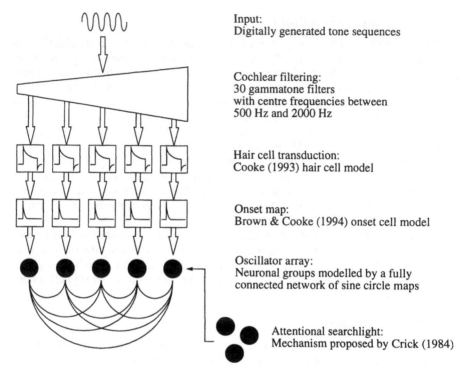

Input:
Digitally generated tone sequences

Cochlear filtering:
30 gammatone filters
with centre frequencies between
500 Hz and 2000 Hz

Hair cell transduction:
Cooke (1993) hair cell model

Onset map:
Brown & Cooke (1994) onset cell model

Oscillator array:
Neuronal groups modelled by a fully
connected network of sine circle maps

Attentional searchlight:
Mechanism proposed by Crick (1984)

FIG. 7.1. Schematic diagram of the model. Note that the attentional searchlight is not explicitly modeled in the current computer simulation.

7.2.3 Neural Network

In the current model, we assume that source segregation is achieved by selective attention to the neural activity in groups of auditory filter channels. The grouping of channels is indicated by the pattern of temporal synchronization in a neural network, consisting of a fully connected array of sine circle maps (Bauer & Martienssen, 1991). Each circle map models the phase dynamics of a neural oscillator, which signals the grouping between its corresponding auditory filter channel and other channels. For clarity, we refer to units in the network as neurons, although each is intended to reflect the activity of a *group* of neurons rather than a single cell. The topology of the network is shown schematically in Figure 7.1. The sine circle map $\varphi(x)$ is defined by

$$\varphi(x) = x + \Omega + \frac{k}{2\pi}\sin(2\pi x) + \eta \,(\mathrm{mod}\,1) \qquad \text{EQ. 7.3}$$

where η is a noise term representing equally distributed random numbers in the range $[0, 10^9]$. The new phase $\theta_i(t + 1)$ of the neuron for channel i is calculated by applying the circle map φ on the old phase $\theta_i(t)$ and on an input value v_i, weighted with a coupling strength κ.

$$\theta_i(t+1) = \frac{1}{1+\kappa}[\varphi(\theta_i(t)) + \kappa\varphi(v_i(t))] \qquad \text{EQ. 7.4}$$

The input value v_i is related to the phases of the other neurons in the network, weighted with a coupling matrix W:

$$v_i(t) = \frac{\sum_j W_{ij}\theta_j}{\sum_j W_{ij}} \qquad \text{EQ. 7.5}$$

For the parameter set used here ($\kappa = 1.5$, $\Omega = 0.618$, $k = 5.0$), neurons in the network exhibit chaotic oscillations. When the coupling strength W_{ij} between two neurons i and j is high, the cells show an identical (but still chaotic) phase response. In contrast, neurons that are weakly coupled give rise to uncorrelated time series.

This behavior is shown in Figure 7.2, where coupling strengths have been manually set to illustrate grouping by phase response $\theta(t)$ in a network of 8 neurons. Neurons in the upper and lower groups of 4 cells are tightly coupled, but there is no coupling between cells of different groups. Although strongly coupled neurons show the same chaotic time series, the noise in the system causes the two groups to decorrelate after a few time steps.

Learning in our network proceeds during simulation as follows. Initially, all neurons are strongly coupled ($W_{ij} = 1$, for all $i{\neq}j$) and therefore show an identical phase response. Subsequently, coupling strengths are modified every 1 ms according to the following learning rule:

$$W_{ij}(t+1) = \Phi(1 - \lambda[1 - W_{ij}(t)] - \gamma|s_i(t) - s_j(t)|) \qquad \text{EQ. 7.6}$$

Here, γ determines the learning speed, $s_i(t)$ is the firing activity of the onset cell for auditory filter channel i and λ determines the rate at which W_{ij} returns to its resting level (unity). The function $\Phi(x)$, given by

$$\Phi(x) = \begin{cases} x & 0 \leq x \leq 1 \\ 1 & x > 1 \\ 0 & x < 0 \end{cases} \qquad \text{EQ. 7.7}$$

confines the value of the coupling strength to the range $[0,1]$. Parameter values were set by inspection ($\lambda = 0.9995$, $\gamma = 0.2$).

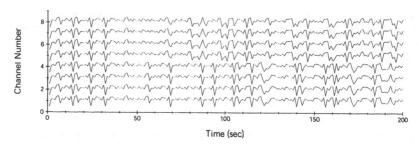

FIG. 7.2. Temporal response in a network of 8 neural oscillators. Neurons in the upper and lower groups of 4 neurons are strongly coupled, but there is no coupling between neurons of different groups.

The learning rule defined by equation 7.6 embodies a principle that is closely related to the Gestalt principle of *common fate*. The principle of common fate describes the tendency to group sensory elements that change in the same way at the same time (Koffka, 1936). Similarly, our learning rule ensures that the coupling is reduced between channels whose onset cells do not exhibit the same degree of activity at the same time.

7.2.4 Attentional Searchlight

The final stage of our model is an attentional mechanism, inspired by Crick's (1984) suggestion that an attentional "searchlight" is located in the thalamus. In physiological terms, the searchlight is expressed by the production of rapid bursts of firing in a subset of thalamic cells. When the thalamic neurons fire in synchrony with the oscillations of a neuronal group (or an assembly of neuronal groups), that group becomes the attentional "foreground" and other neuronal groups are relegated to the "background". The action of this mechanism is rather similar to that of a stroboscope: by firing in synchrony with the oscillations of a particular neuronal group, the thalamic searchlight is able to isolate it from other neural activity.

Although an attentional searchlight is proposed as part of the scheme described here, an attentional mechanism is not currently implemented in our computer model (this is ongoing work). Rather, we consider the temporal correlation between the activity of pairs of neurons in the oscillator network. The correlation between two time series $X(t)$ and $Y(t)$ is given by

$$C = \frac{\sum x(t)y(t)}{\sqrt{\sum x^2(t) \sum y^2(t)}}$$

EQ. 7.8

where $x(t) = X(t) - \langle X \rangle$ and $y(t) = Y(t) - \langle Y \rangle$. If two neurons in our network are strongly coupled, the temporal correlation C of their responses will be high. Consequently, an attentional searchlight that is synchronized with the activity of one of the neurons is inevitably synchronized with the other; the two cells form a group. Similarly, the temporal correlation C will be low between two neurons that are weakly coupled. In this case, the attentional searchlight may synchronize with the activity of one cell or the other, but not both simultaneously. Hence, in Bregman's (1990) terms, high temporal correlation between neurons indicates "temporal coherence" and low correlation indicates "streaming". Section 7.3.1 of the chapter discusses this mechanism in greater detail.

7.3 SIMULATION RESULTS

This section aims to illustrate the properties of the model by examining its ability to replicate a number of two-tone streaming phenomena reported in the psychophysical literature. All simulations used the same parameter set.

7.3.1 Grouping by Temporal Proximity and Frequency Proximity

A number of workers have studied auditory stream segregation by presenting listeners with a repeating sequence of tones that alternate between two different frequencies (e.g., McAdams & Bregman, 1979; Van Noorden, 1975). At long tone repetition times (the time interval between the onset of consecutive tones) listeners tend to hear *temporal coherence*; the sequence is perceived as a single tone that changes in frequency, with a period corresponding to the tone repetition time. When the tone repetition time is short, *streaming* occurs; listeners hear two alternating tones with a period corresponding to twice the tone repetition time. In the latter condition, listeners are able to attend either to the high tones or the low tones in the sequence, but not to both. The perception of auditory streams in alternating tone sequences is similarly influenced by the frequency separation between the two tones. In summary, streaming is more likely to occur if the tone repetition time is short and the frequency separation between the two tones is large.

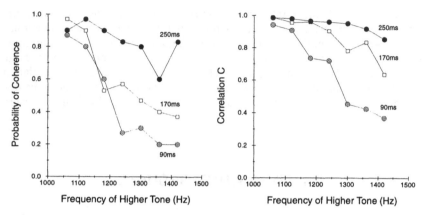

FIG. 7.3. Comparison of subject results (left) and model simulation (right) for the two-tone streaming study of Beauvois and Meddis (1991). For clarity, only the 90 ms, 170 ms and 250 ms conditions are shown.

Here, we consider a two-tone streaming study reported by Beauvois and Meddis (1991). They presented listeners with sequences of alternating 40 ms tones, with repetition times between 50 ms and 250 ms. The frequency of the lower tone was always 1000 Hz, whereas the frequency of the higher tone was varied between 1060 Hz and 1420 Hz. After a 15 second presentation of a sequence, listeners were asked to indicate whether the last percept they heard was streaming or temporal coherence. The pooled responses for six subjects are shown in the left panel of Figure 7.3. As expected, listeners were more likely to report streaming if the sequence had a short tone repetition time and a large frequency separation.

The right panel of Figure 7.3 shows the output generated by our model for the stimuli used by Beauvois and Meddis. Temporal correlation C was computed between the phase response of the two neurons in the network whose corresponding auditory filter center frequencies were closest to the frequencies of the two tones. The correlation values shown were averaged over the last second of a 15 second stimulus. Clearly, the model exhibits the required behavior; temporal correlation between the two neurons decreases as the frequency separation increases, and this effect is exaggerated at short tone repetition times.

The model displays a sensitivity to tone repetition time because temporal integration is inherent in the learning rule for the oscillator network (see equation 7.6). Recall that if two neurons in the network receive different levels of input activity, the coupling strength W between them is reduced. In the absence of further stimulation, W returns exponentially to its resting level at a rate determined by λ. However, the stimuli used here consist of repeating tones, so W is only able to recover in the time between the onset of successive tones. It follows that the coupling strength between two neurons will be driven lower (and hence the temporal correlation between the neurons will be lower) at short tone repetition times, since there is only a brief period in which W can recover.

The sensitivity of the model to frequency separation is due principally to the tuning properties of the auditory filterbank. For example, consider an auditory filter with a center frequency close to the frequency of the lower tone in the sequence. When the frequency separation between the higher and lower tones is small, the higher tone will generate a response in this filter that is almost equivalent to the response produced by the lower tone. Hence, the term $|s_i - s_j|$ in equation 7.6 is small, and the coupling strength W_{ij} between the low tone and high tone channels in the oscillator network remains high. However, if the frequency separation is large, the higher tone will generate a much smaller response in the auditory filter centered on the lower tone. Consequently, $|s_i - s_j|$ is large, and the coupling strength W_{ij} is weakened, causing the neural oscillators for the high tone and low tone channels to decorrelate.

Bregman's (1990) account of auditory scene analysis uses Gestalt perceptual grouping principles to explain stream formation. Specifically, it attributes the effects of tone repetition time and frequency separation described above to the action of the Gestalt principles of *temporal proximity* and *frequency proximity*. In contrast, our model adopts a lower level of explanation; grouping by temporal proximity and frequency proximity are seen as emergent properties of the auditory periphery and a simple neural network.

Van Noorden (1975) noted the existence of two perceptual boundaries which depend upon the repetition time and frequency separation in a repeating two-tone sequence (see also McAdams & Bregman, 1979). Subjects always hear a single stream below the *fission boundary* and always hear two streams above the *temporal coherence boundary*. Between these two boundaries is a perceptually ambiguous region, in which listeners can either hear streaming or temporal coherence depending on their attentional set.

In our model, the existence of these boundaries can be explained in terms of the degree of tolerance allowed when the attentional searchlight synchronizes with neuronal groups. We assume that if the temporal correlation C between two neurons in the oscillator network exceeds a critical value, the attentional searchlight is able to synchronize to both neurons and they form a group. Similarly, if the correlation C between two neurons falls below a critical value, the attentional searchlight is able to synchronize with one neuron or the other, but not both. These upper and lower correlation thresholds correspond to the fission boundary and temporal coherence boundary respectively. At intermediate correlation values, we assume that the attentional searchlight may synchronize with one neuron or both neurons; perceptual ambiguity occurs because of the tendency of the searchlight to drift randomly between these two organizations.

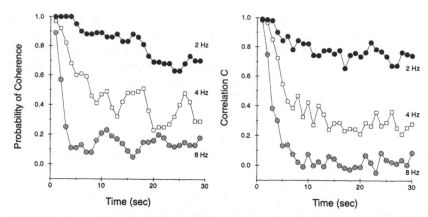

FIG. 7.4. Comparison of subject results (left) and model simulation (right) for the square wave FM stimuli used by Anstis and Saida (1985).

A final point concerns the observation of Dannenbring and Bregman (1976) that two-tone streaming is unaffected by the length of the silence between successive tones, provided that the time interval between tone onsets remains constant. Our model is compatible with this finding, because the coupling strengths in the oscillator network are modified according to the activity of onset cells, which fire only at the start of an applied stimulus.

7.3.2 Build-up of Streaming Over Time and Good Continuation

Anstis and Saida (1985) studied the relationship between repetition rate and probability of coherence over the total duration of a stimulus. Their findings are shown in the left panel of Figure 7.4. Subjects were presented with square wave frequency modulated (FM) signals at alternation rates of 2 Hz, 4 Hz and 8 Hz, and asked to continuously monitor the number of streams that they heard. Stimuli were 30 seconds long, and alternated between frequencies of 800 Hz and 1200 Hz. The probability that subjects heard a single coherent stream was high at the start of each stimulus, but became progressively less likely over time. Further, the build-up of streaming occurred more rapidly at faster alternation rates.

The response of our model to Anstis and Saida's stimuli is shown in the right panel of Figure 7.4. Average temporal correlation C was computed every second between the two neurons in the oscillator network whose corresponding filter center frequencies were closest to 800 Hz and 1200 Hz. Clearly, the model shows the required trend; temporal correlation (and hence temporal coherence) falls off over time at a rate which is determined by the alternation rate.

Anstis and Saida also presented their subjects with sinusoidal FM stimuli at the same alternation rates. Their results are shown in the left panel of Figure 7.5. Again, probability of coherence tended to fall off over time, but the

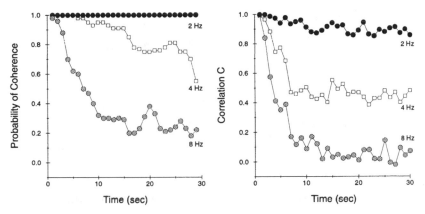

FIG. 7.5. Comparison of subject results (left) and model simulation (right) for the sinusoidal FM stimuli used by Anstis and Saida (1985).

sinusoidally modulated stimuli sounded more coherent than the square wave modulated stimuli. The model simulation, shown in the right panel of Figure 7.5, is consistent with this finding.

   An alternative representation of Anstis and Saida's data is shown in the left panel of Figure 7.6. Here, the curves in the left panels of Figures 7.4 and 7.5 have been reduced to a single point by integration, giving a mean probability of temporal coherence over the duration of the stimulus. This representation of the data indicates that the probability of hearing a single coherent stream declines as a linear function of log alternation rate. Note that the points for square wave FM are lower than those for sinusoidal FM, indicating that the latter gave more coherence. The right panel of Figure 7.6 shows the results of our simulation in the same form. Clearly, the model provides a good match to the psychophysical data.

   An auditory scene analysis account would hold that sinusoidal FM gives more coherence than square wave FM because the glides connecting the low tones and high tones invoke a Gestalt principle of *good continuation*. In contrast, our model explains this effect in terms of the bandpass properties of auditory filters. An auditory filter with a center frequency close to the higher or lower component of a square wave FM stimulus will respond strongly for half a cycle of the stimulus. In contrast, the filter will only respond strongly to a sinusoidal FM stimulus during the period in which the signal sweeps through the filter's pass band. Consequently, a sinusoidal FM stimulus will elicit a smaller response in our onset map than a square wave FM stimulus, and the temporal correlation between neurons in the oscillator network will be higher (more coherent) for sinusoidal FM and lower (less coherent) for square wave FM.

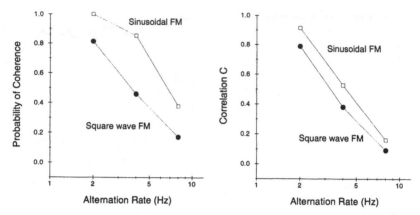

FIG. 7.6. Data from Figures 7.4 and 7.5, plotted to show that temporal coherence falls linearly with increasing log alternation rate. Subject responses are shown on the left, the model simulation is shown on the right.

7.3.3 Grouping by Common Onset

Bregman and Rudnicky investigated whether two tones that overlap in time can be heard as separate streams, rather than as a single complex tone (Bregman, 1990). They presented listeners with a repeating cycle of two tones, with a fixed duration (250 ms) and repetition rate. The percentage of the length of one tone that overlapped the other tone was varied, as shown in the left panel of Figure 7.7. Listeners reported that the two tones fused into a single complex when the temporal overlap was greater than 88% (approximately synchronous onsets). Conversely, sequences that were only 50% overlapped segregated into separate streams.

The response of our model to Bregman and Rudnicky's stimulus, shown in the right panel of Figure 7.7, is compatible with this finding. Temporal correlation C, averaged over the last second of a 15 second stimulus, is high when the percentage overlap is high, indicating a coherent percept. However, temporal correlation falls off as the percentage overlap decreases, indicating a streaming percept. The sensitivity of our model to onset synchrony follows directly from the learning rule given in equation 7.6, which reduces the coupling strength between neural oscillators that do not receive the same level of input activity at the same time.

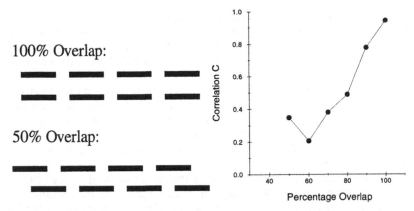

100% Overlap:

50% Overlap:

FIG. 7.7. Model simulation (right) of the Bregman and Rudnicky streaming study. The form of the stimulus, a repeating cycle of overlapping tones, is shown on the left.

7.4 DISCUSSION

This chapter describes a model of primitive auditory stream segregation in which the grouping of peripheral channels is indicated by temporal synchronization in a network of neural oscillators. Despite its simplicity, the model has considerable explanatory power and gives a qualitative match to data from a number of psychoacoustical studies.

A notable feature of the model is that all neural oscillators are tightly coupled in their resting state. This implies that the default condition of organization in the model is perceptual fusion; all components of a stimulus are assumed to have originated from the same environmental source unless there is evidence for segregating them. Indeed, there is good reason to believe that fusion is the default condition of auditory organization; for example, a long burst of white noise contains random cues for fusion and segregation, but is perceived as a single coherent sound (Bregman, 1990). Similarly, it follows from the learning rule in equation 7.6 that, in response to a white noise stimulus, the network of neural oscillators in our model would remain strongly coupled.

A number of other workers proposed models of auditory stream segregation. The model of Beauvois and Meddis (1991) contains elements that are superficially similar to those of our own model; an auditory filterbank, random noise and a temporal integration mechanism. In their scheme, temporal integration is implemented as a leaky integration of auditory filter outputs. In contrast, our model incorporates temporal integration in a learning rule which governs the modification of coupling strengths in a neural network. Similarly, the function of random noise differs in the two models. Beauvois and Meddis' model uses noise to cause random changes of focus in a low-level attentional mechanism. The function of noise in our system is to ensure that uncoupled neural oscillators show a different phase response; attention is attributed to a higher-level "neural searchlight".

A final point is that our model has a rather wider applicability than the Beauvois and Meddis (1991) scheme. Section 7.3.3 demonstrated that our model is able to account for certain streaming phenomena in which frequency components overlap in time, so-called *simultaneous* grouping (Bregman, 1990). In contrast, the model of Beauvois and Meddis only addresses the *sequential* grouping of frequency components.

Our model bears a closer resemblance to the recent work of Wang (1996), which also uses temporal synchronization to indicate grouping in a network of neural oscillators. Wang's auditory model is a development of an earlier model of visual perception (Wang, 1993) and hence treats auditory stimuli very much like visual stimuli; the input to his auditory model is a symbolic time-frequency pattern. In contrast, our model processes sampled acoustic waveforms, and incorporates a realistic simulation of the auditory periphery. Our approach is arguably more elegant. For example, Wang's network is sensitive to frequency proximity because the coupling strengths between cells in his neural network are "hard-wired" to give the appropriate behavior. There is no such hard-wiring in our network; rather, sensitivity to frequency proximity emerges from the bandpass properties of the auditory filterbank.

The majority of neural oscillator models described in the literature (e.g., König & Schillen, 1991) exhibit periodic oscillations. However, it has been suggested that the number of phases available in networks of such neurons is significantly less than the minimum required to perform useful computations (Cairns, Baddeley & Smith, 1993). In contrast, the sine circle maps employed here exhibit chaotic oscillations, allowing a large number of groups to be distinguished that show no correlation between members of different groups but perfect correlation between members of the same group. In practice, the number of different groups that the network can sustain is limited by the length of the temporal correlation window and the resolution of the phase variable θ (Bauer & Martienssen, 1991).

The model is presented here principally as proof of concept. Clearly, it would be desirable to incorporate a wider range of auditory features in the model, and these could be provided by the feature detecting maps proposed in Brown and Cooke (1994). In particular, the addition of a mechanism for grouping harmonically related components would allow us to simulate a larger number of psychoacoustic streaming studies. Additionally, in its present form the model only addresses primitive (bottom up) auditory grouping. We are currently investigating the application of schema-driven (top down) grouping mechanisms in the model. It is anticipated that such mechanisms will allow the simulation of perceptual restoration phenomena (see also Cooke & Brown,

1993) and will relate closely to our work on the recognition of partially occluded speech signals (Green, Cooke & Crawford, 1995).

ACKNOWLEDGMENTS

The authors would like to thank Darryl Godsmark for his suggestions on the form of the learning rule. Thanks also to Andrew Morris for his comments on an earlier draft of this paper. This research is supported by SERC Image Interpretation Initiative Grant GR/H53174, EPSRC Standard Research Grant GR/K18962 and a grant from the Nuffield Foundation.

REFERENCES

Anstis, S. & Saida, S. (1985). Adaptation to auditory streaming of frequency modulated tones. *Journal of Experimental Psychology: Human Perception and Performance*, 11, 257-271.

Bauer, M. & Martienssen, W. (1991). Coupled circle maps as a tool to model synchronization in neural networks. *Network*, 2, 345-351.

Beauvois, M. W. & Meddis, R. (1991). A computer model of auditory stream segregation. *Quarterly Journal of Experimental Psychology*, 43A, 517-541.

Bregman, A. S. (1990). *Auditory Scene Analysis*. Cambridge, MA: MIT Press.

Brown, G. J. & Cooke, M. P. (1994). Computational auditory scene analysis. *Computer Speech and Language*, 8, 297-336.

Cairns, D. E., Baddeley, R. J., & Smith, L. S. (1993). Constraints on synchronising oscillator networks. *Neural Computation*, 5, 260-266.

Cooke, M. P. (1993). *Modelling auditory processing and organisation*. Cambridge, UK: Cambridge University Press.

Cooke, M. P. & Brown, G.J. (1993). Computational auditory scene analysis: Exploiting principles of perceived continuity. *Speech Communication*, 13, 391-399.

Crick, F. (1984). Function of the thalamic reticular complex: The searchlight hypothesis. *Proceedings of the National Academy of Sciences USA*, 81, 4586-4590.

Dannenbring, R. & Bregman, A. S. (1976). Effect of silence between tones on auditory stream segregation. *Journal of the Acoustical Society of America*, 59, 987-989.

Glasberg, B. & Moore, B. C. J. (1990). Derivation of auditory filter shapes from notched-noise data. *Hearing Research*, 47, 103-138.

Gray, C. M., König, P., Engel, A. K., & Singer, W. (1989). Oscillatory responses in cat visual cortex exhibit inter-columnar synchronization which reflects global stimulus properties. *Nature*, 338, 334-337.

Green, P. D., Cooke, M. P., & Crawford, M. D. (1995). Auditory scene analysis and Hidden Markov Model recognition of speech in noise. *Proceedings of the IEEE International Conference on Acoustics, Speech and Signal Processing*, pp. 401-404.

Koffka, K. (1936). *Principles of Gestalt psychology*. New York: Harcourt and Brace.

König, P. & Schillen, T. B. (1991). Stimulus-dependent assembly formation of oscillatory responses: I. Synchronization. *Neural Computation*, 3, 155-166,

von der Malsburg, C. & Schneider, W. (1986). A neural cocktail party processor. *Biological Cybernetics*, 54, 29-40.

McAdams, S. & Bregman, A. S. (1979). Hearing musical streams. *Computer Music Journal*, 3, 26-43.

Moore, D. R. (1987). Physiology of higher auditory system. *British Medical Bulletin*, 43, 856-870.

Patterson, R. D., Holdsworth, J., Nimmo-Smith, I., & Rice, P. (1988). *Implementing a gammatone filterbank*. Cambridge, UK: MRC Applied Psychology Unit Report 2341.

Shofner, W. P. & Young, E. D. (1985). Excitatory/inhibitory response types in the cochlear nucleus: Relationships to discharge patterns and responses to electrical stimulation of the auditory nerve. *Journal of Neurophysiology*, 54, 917-939.

Van Noorden, L. P. A. S. (1975). *Temporal coherence in the perception of tone sequences*. Unpublished doctoral dissertation, Institute for Perception Research, Einthoven University.

Wang, D. (1993). Modelling global synchrony in the visual cortex by locally coupled neural oscillators. In *Proceedings of the 15th Annual Conference of the Cognitive Science Society*. Hillsdale, NJ: Erlbaum, pp. 1058-1063.

Wang, D. (1996). Primitive auditory segregation based on oscillatory correlation. *Cognitive Science*, 20, 409-456.

8

The IPUS Blackboard Architecture as a Framework for Computational Auditory Scene Analysis

Frank Klassner and Victor Lesser
University of Massachusetts

S. Hamid Nawab
Boston University

The Integrated Processing and Understanding of Signals (IPUS) architecture is designed for complex environments, which are characterized by variable signal to noise ratios, unpredictable source behaviors, and the simultaneous occurrence of objects whose signal signatures can distort each other. Because auditory scene analysis is replete with issues concerning the relationship between SPA-appropriateness and multi-sound interactions in complex environments, much of our experimental work with IPUS has focused on applying the architecture to this problem. In this chapter we present our work-in-progress in scaling-up our IPUS sound understanding testbed to accommodate a library of 40 sounds covering a range of types (e.g., impulsive, harmonic, periodic, chirps) and to analyze scenarios with three or four sounds.

8.1 INTRODUCTION

In previous articles (Lesser et al., 1991; Lesser et al., 1993; Lesser et al., 1995) we discussed the Integrated Processing and Understanding of Signals (IPUS) architecture as a general framework for structuring bidirectional interaction between front-end signal processing algorithms (SPAs) and signal understanding processes. This architecture is designed for complex environments, which are characterized by variable signal to noise ratios, unpredictable source behaviors, and the simultaneous occurrence of objects whose signal signatures can distort each other. In these environments, the choice of numeric signal processing algorithms (SPAs) and their control parameter values is crucial to the generation of evidence for symbolic interpretation processes. Parameter values inappropriate to the current scenario can render an interpretation system unable to recognize entire classes of signals. We designed IPUS to provide an interpretation system with the ability to dynamically modify its front-end SPAs to handle scenario changes and to reprocess ambiguous or distorted data. This adaptation is organized as two concurrent search processes: one for correct interpretations of SPAs' outputs and another for SPAs and control parameters appropriate for the environment. Interaction between these search processes is structured by a formal theory of how inappropriate SPA usage can distort SPA output.

Because auditory scene analysis is replete with issues concerning the relationship between SPA-appropriateness and multisound interactions in complex environments, much of our experimental work with IPUS has focused on applying the architecture to this problem. Our earlier work focused on the first version of the IPUS sound understanding testbed (configuration C.1) and dealt with evaluating how well IPUS could use small libraries of sound models (5 to 8) and small sets of signal processing algorithms (SPAs) to analyze acoustic scenarios with two or three sounds. In this chapter we first summarize the IPUS approach and then present our ongoing work in developing the C.2 testbed. The work on the C.2 configuration is intended to investigate how to practically scale-up the IPUS testbed to accommodate a large library of 40 sounds covering a range of types (e.g., impulsive, harmonic, chirps, periodic) (Klassner et al., 1995; Klassner, 1996) and to analyze scenarios with three or four sounds. In particular, we (1) present the evidential hierarchy we use to describe acoustic sources' features and (2) describe the incorporation of *approximate processing* techniques that compute or analyze only subregions of an acoustic scenario's spectrogram based on high-level source models and expectations that arise from the scenario's emerging interpretation.

8.2 IPUS OVERVIEW

IPUS uses an iterative process for converging to the appropriate SPAs and interpretations for a signal. The following discussion summarizes the IPUS blackboard architecture shown in Figure 8.1. For each block of data, the loop starts by processing the signal with an initial configuration of SPAs. These

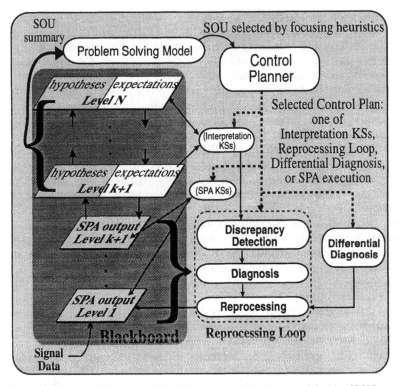

FIG. 8.1. Abstract Architecture. The abstract architecture used in the IPUS sound understanding testbed.

SPAs are selected not only to identify and track the signals most likely to occur in the environment, but also to provide indications of when less likely or unknown signals have occurred. In the next part of the loop, a *discrepancy detection* process tests for discrepancies between the output of each SPA in the current configuration and (1) the output of other SPAs in the configuration, (2) application-domain constraints, and (3) the output's anticipated form based on high-level expectations. Architectural control permits this process to execute both after SPA output is generated and after interpretation problem solving hypotheses are generated. If discrepancies are detected, a *diagnosis* process attempts to explain them by mapping them to a sequence of qualitative distortion hypotheses. The loop ends with a *signal reprocessing* stage that proposes and executes a search plan to find a new front-end (i.e., a set of instantiated SPAs) to eliminate or reduce the hypothesized distortions. After the loop's completion for a given data block, if there are any similarly-rated competing top-level interpretations, a *differential diagnosis* process selects and executes a reprocessing plan to find outputs for features that will discriminate among the alternatives.

Although the architecture requires the initial processing of data one block at a time, the loop's diagnosis, reprocessing, and differential diagnosis components are not restricted to examining only the current block's processing results. If the current block's processing results imply the possibility that earlier blocks were misinterpreted or inappropriately reprocessed, those components can be applied to the earlier blocks as well as the current blocks. Additionally, reprocessing strategies and discrepancy detection application-constraints tests can include the postponement of reprocessing or discrepancy declarations until specified conditions are met in the next data block(s).

The philosophy behind the IPUS architecture recognizes that different signal representations, with different levels of precision, are required for interpreting complex signals such as auditory signals as they change over time. There is no single fixed processing strategy (sequence of fixed-parameter SPAs) that will provide adequate evidence for all sounds under all scenarios. Given the dynamic nature of many environments, it is inefficient to predetermine complete processing strategies to provide precise evidence for all possible interactions among all possible sounds. A fixed front-end designed to generate well-resolved tracks, for instance, would needlessly perform costly high-resolution time-frequency analysis even when only a single sound with widely-separated tracks is present in the acoustic signal.

IPUS permits interpretation system designers to specify a wide range of signal features and specialized SPAs for detecting them, without requiring detailed strategies for applying the SPAs with particular control parameter values. The architecture is designed to use the signal-processing theory underlying the SPAs to opportunistically select signal features found in a preliminary analysis to serve (1) as the basis for further examination or (2) as the basis for decisions to apply more expensive, specialized SPAs. The precision of the output from the SPAs can vary according to the system's time constraints and the complexity of the signal's current partial interpretation.

At this point one might question these criticisms of fixed, one-pass front ends by claiming that the human auditory system, with its cochlear signal processing, is an example of just such a system that handles a wide variety of complex acoustic scenarios quite well. Although the human auditory system's "hardware" may indeed be fixed, there is evidence that the system's cognitive component is not. The set of highlevel features and their required precisions that the cognitive components use to identify sounds (e.g. duration, synchronization, expectations from sequentiality) changes frequently as the auditory system interprets real-world signals.

Ellis (Ellis, 1996) summarizes work (McAdams, 1984; Warren, 1984) that supports this claim of shifting feature selection in the human auditory system. The McAdams work reported that during the initial presentation of an oboe note to human observers, a single note is perceived. However, as the note progresses, the even harmonics of the note undergo progressively deeper frequency modulation. This led observers to change their initial interpretation of the spectral energy as a single note to one in which there were two distinct sounds, and to apply that changed interpretation to the entire note. The Warren work examined how changes in the order of presentation of alternating wide (0-2KHz) and narrow (0-1KHz) noise bands influenced the grouping that humans performed on the signal components. When the narrow band came first, observers interpreted the signal as containing a continuous 0-1KHz sound with a periodic 1KHz-2KHz sound. When the wide band started the alternation, observers reported that the first 0-2KHz band was initially interpreted as a single sound, but that as the alternation progressed, the interpretation of the first band was dropped in favor of combining it with the rest of the signal and obtaining the same interpretation as that found for the first alternation.

8.3 NEW TESTBED EVIDENCE HIERARCHY

This section discusses the set of signal feature representations we use in the enhanced testbed. Our extended testbed uses thirteen partially ordered evidence representations to construct an interpretation of incoming signals. They are implemented through thirteen levels on the architecture's hypothesis blackboard. Figure 8.2 illustrates the support relationships among the representations, while the following discussion highlights the representations' content:

(1) The first blackboard level is simply the raw waveform data. This representation is required in spite of its space requirement since the testbed architecture will sometimes need to reprocess data. To conserve space, only the last 3 seconds of waveform data are kept on the testbed's blackboard.

(2) The second level contains hypotheses on the envelope, or shape, of the time-domain signal. These hypotheses also maintain statistics such as zero-crossing density and average energy for each block of signal data. This is a new representation in the C.2 configuration that was not in the C.1 configuration.

(3) The third level contains spectral hypotheses derived for each waveform segment through algorithms such as the Short-Time Fourier Transform (STFT) and the Wigner Distribution (Claasen & Meclenbrauker, 1980). One of the SPAs for producing hypotheses on this level, the Quantized STFT, is new to the C.2 configuration and is discussed in the next section.

(4) The fourth level contains peak hypotheses derived for each spectrum. These are used to indicate narrow-band features in a signal's spectral representation.

(5) The fifth level contains energy-shift hypotheses, which indicate sudden energy changes in the time-domain envelope. This is a new representation.

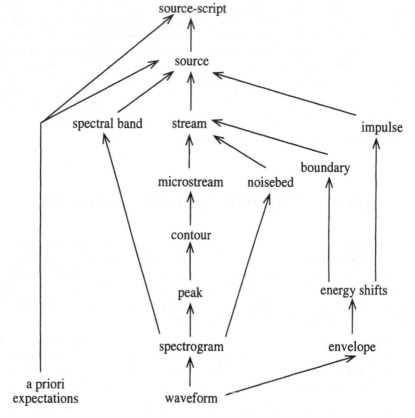

FIG. 8.2. Distortion Example. The acoustic abstraction hierarchy for the extended IPUS testbed.

(6) The sixth level contains time-domain event hypotheses, which group shifts into boundaries (i.e., a step-up or step-down in time domain energy indicating the start or end of some sound) and impulses (i.e., sudden spikes in the signal). This is also a new representation.

(7) The seventh level contains of contour hypotheses, each of which corresponds to a group of peaks whose time indices, frequencies, and amplitudes represent a contour in the time-frequency-energy space with uniform frequency and energy behavior.

(8) The eighth level contains spectral band hypotheses, which identify regions of activity in spectrograms from the third level.

(9) The ninth level contains microstream hypotheses supported by one contour or a sequence of contours. Each microstream has an energy pattern consisting of an attack region (signal onset), a steady region, and a decay (signal fadeout) region.

(10) The tenth level contains noisebed hypotheses supported by regions within spectra. Noisebeds represent the wideband component of a sound source's acoustic signature. Microstreams often form "ridges" on top of noisebed "plateaux," but not every noisebed has an associated microstream.

(11) On the eleventh level we apply perceptual streaming criteria developed in the psychoacoustic research community (Bregman 1990) to group microstreams and noisebeds as support for stream hypotheses, or entities to be recognized as sound sources. Specifically, our testbed knowledge sources group microstreams together when they have similar fates (e.g., synchronized onset- and end-times, synchronized chirp behavior), or when they share a harmonic relationship. Noisebeds are predicted and searched for only after a stream has been identified as a particular source's signature.

(12) At the twelfth level, stream hypotheses, with their durations supported by boundaries, are interpreted as sound-source hypotheses according to how closely they match source-models available in the testbed library. Partial matches (e.g., a stream missing a microstream, or a stream with duration shorter than expected for a particular source) are accepted and posted, but these are penalized with uncertainty (referred to as SOU-Source Of Uncertainty, in the architecture diagram). These uncertain hypotheses will later cause the testbed to attempt to account for the missing or ill-formed evidence (e.g., microstreams or noisebeds) as artifacts of improper front-end processing.

(13) The thirteenth level contains sound script hypotheses, which represent hypotheses about the temporal streaming of a sequence of sources into a single unit (e.g., a periodic source such as footsteps being composed of a sequence of evenly-spaced footfalls, or the combination of cuckoo-chirps and bell-tones in a cuckoo-clock chime). This is a new representation in the C.2 configuration.

Note that only a subset of the features in the hierarchy are sought for at any given time in the system's execution. For each representation, the testbed has several possible SPAs or interpretation knowledge sources (KSs) that can produce it, with varying degrees of precision.

As mentioned earlier, not all scenarios require precise feature hypotheses. In the next section we present two techniques the C.2 testbed uses to produce approximate hypotheses that help to limit the architecture's search for signal interpretations while reducing the time spent by the testbed's front-end on initial analysis of the signal.

8.4 TESTBED APPROXIMATE PROCESSING TECHNIQUES

Approximate processing (Decker et al., 1990) refers to the concept of deliberately limiting search processes in order to trade off certainty for reduced execution time. We have found two classes of approximation techniques particularly useful in reducing search-time for interpretations of acoustic scenarios: *Data Approximation* limits the characteristics of the data to be inspected by the search process and consequently results in solutions that are less precise and more uncertain; *Knowledge Approximation* eliminates or simplifies the constraints utilized by the search process. In this case, certainty in the answer obtained is reduced because it may be different from the answer obtained with the full set of constraints. We use several versions of an STFT approximation algorithm (Nawab & Dorken, 1995) that sacrifices certainty (e.g., frequency resolution, time resolution, or spectral coverage) in its output data in exchange for a reduction in processing time by an order of magnitude below the precise STFT's requirements. This reliance on data approximation permits the formulation of interpretation strategies that save time by first obtaining a rough approximation of a scenario's spectrogram and then refining only those portions of it that remain ambiguous when compared with high-level sound models.

Our introduction of the *spectral band* abstraction level represents a knowledge approximation technique that avoids over-reliance on strict narrowband descriptions of sounds by mapping rough clusters of spectral activity in a spectrogram to only those sounds in the sound library that overlap those frequency regions. By abandoning unrestricted bottom-up search for narrowband components in favor of selective search in subregions of spectrograms indicated by source models, the testbed avoids effort in verifying improbable sounds' tracks.

As an example of how computational verification effort can be reduced, consider the following testbed behavior combining approximate and precise computation of the STFT. When the testbed is directed to use minimal time for front-end processing, and it selects from its SPA database an approximate STFT SPA instance that meets the time requirements. This SPA is applied as part of the front-end processing that produces spectral-band hypotheses about where spectral energy is concentrated in the scenario spectrogram. From these spectral bands the testbed's interpretation KSs hypothesize the existence of several sounds, some, or all of which may be simultaneously present. In the cases where two or more sounds' narrowband tracks cover the same frequency region, the testbed uses differential diagnosis to generate a reprocessing strategy that uses a pruned, precise STFT SPA to provide high-resolution spectral data *only for the ambiguous frequency region*, saving verification (and front-end processing) effort.

8.5 CONCLUSION

In addition to support for designing adaptive, low-cost front-ends, IPUS offers a framework for integrating top-down, expectation-driven processing with bottom-up, psychoacoustically oriented processing. The architecture unifies SPA reconfiguration performed for symbolic-based interpretation processes with that performed for numeric-based processes as a single reprocessing concept controlled by the presence of various classes of uncertainty. For example, the same contouring algorithm is executed with different parameters depending on whether an expectation exists for any narrowband tracks in the spectrogram region in which it is applied. If there are model-based expectations with little or no uncertainty, focused contouring that relies on the frequency tracks' properties is performed, otherwise bottom-up contouring is performed with parameters set to detect steady-state behavior of tracks that may be originating from sounds that have no models in the testbed's sound library. Another example concerns the expectation-driven parameter adaptation that IPUS performs when partial evidence for a sequential stream (e.g., sound script) such as a series of footsteps or phone rings is available. If it is possible that an expected sound's tracks will be indistinguishable from those of another sound due to poor frequency resolution afforded by the current parameters of the front-end STFT, the testbed anticipates the distortion and resets the STFT parameters.

In our presentation we discuss the testbed's overall performance as well as the specific recognition improvements obtained by the testbed's new representations and approximate processing techniques.

Information about the further development of this work since its original presentation at the 1995 Computational Auditory Scene Analysis Workshop can be found in Klassner (1996); the IPUS papers cited in this chapter, as well as papers published since the original presentation, are available on the World Wide Web at http://dis.cs.umass.edu/research/ipus/ipus-pubs.html.

8.6 ACKNOWLEDGMENTS

This work was supported by the Office of Naval Research under contract N00014-92-J-1450. The content does not necessarily reflect the position or the policy of the Government, and no official endorsement should be inferred.

REFERENCES

Bregman, A., (1990). *Auditory Scene Analysis: The Perceptual Organization of Sound.* Cambridge, MA: MIT Press.

Claasen, T. & Meclenbrauker, W., (1980). *The Wigner Distribution: A Tool for Time-Frequency Signal Analysis*, (Phillips J. Res.), vol. 35, pp. 276-350.

Decker, K., Lesser, V., & Whitehair, R., (1990) Extending a Blackboard Architecture for Approximate Processing. *Journal of Real-Time Systems*, 2, pp. 47-79, Kluwer Academic Publishers.

Ellis, D. P. W., (1996). *Prediction-Driven Computational Auditory Scene Analysis*, Ph.D. thesis, Dept. of Elec. Eng. & Comp. Sci. Massachusetts Institute of Technology.

Klassner, F., (1996). *Data Reprocessing in Signal Understanding Systems.* Ph.D. thesis, Comp. Sci. Dept., Univ. of Massachusetts, Amherst.

Klassner, F., Govindu, V. M., & Mani, R., (1995). *The IPUS C.2 sound understanding testbed acoustic modeling framework and sound library*, Technical Report 95-65, Comp. Sci. Dept., Univ. of Massachusetts, Amherst.

Lesser, V., Nawab, S. H., & Klassner, F., (1995). IPUS: an architecture for the integrated processing and understanding of signals. *Artificial Intelligence*, vol. 77, no. 1, August.

Lesser, V., Nawab, S. H., Gallastegi, I., & Klassner, F., (1993). IPUS: an architecture for integrated signal processing and signal interpretation in complex environments. *The Proceedings of the 1993 National Conference on Artificial Intelligence* (AAAI-93), pp. 249-255, Washington, DC.

Lesser, V., Nawab, S. H., Bhandaru, M., Cvetanović, Z., Dorken, E., Gallastegi, I., & Klassner, F., (1991). *Integrated Signal Processing and Signal Understanding.* Technical Report 91-34, Comp. Sci. Dept., Univ. of Massachusetts, Amherst.

McAdams, S., (1984). *Spectral Fusion, Spectral Parsing and the Formation of Auditory Images.* Ph.D. thesis, CCRMA, Stanford University.

Nawab, H., & Dorken, E., (1995). Quality versus efficiency tradeoffs in STFT computation, *IEEE Trans. on Signal Processing*.

Warren, R. M., (1984). Perceptual restoration of obliterated sounds. *Psychological Bulletin*, vol. 96, pp. 371-383.

9

Application of the Bayesian Probability Network to Music Scene Analysis

Kunio Kashino
NTT Basic Research Laboratories

Kazuhiro Nakadai, Tomoyoshi Kinoshita, and Hidehiko Tanaka
University of Tokyo

We propose a process model for hierarchical perceptual sound organization that enables the recognition of *perceptual sounds* in incoming sound signals. We consider perceptual sound organization as a scene analysis problem in the auditory domain. Our current application is a *music scene analysis* system that recognizes rhythm, chords, and source-separated musical notes included in incoming music signals. The process model consists of multiple processing modules and a probability network for information integration. Its structure is conceptually based on the blackboard architecture. However, employment of the Bayesian probability network has facilitated integration of multiple sources of information provided by autonomous modules without global control knowledge.

9.1 INTRODUCTION

Humans recognize or understand the existence, localization, and movement of external entities through the five senses. We call this function scene analysis. Scene analysis is viewed here as an information processing that produces valid symbolic representations of external entities or events based on sensory data and stored knowledge. We use the term visual scene analysis for analysis through optical (or visual) information, and the term auditory scene analysis for analysis based on acoustic (or auditory) information. It is widely accepted that research dedicated to the artificial realization of those functions as well as physiological and psychological studies are becoming increasingly important as we enter the multimedia era.

From the engineering point of view, however, the current state of research on auditory scene analysis is still in its infancy when compared with the wide spectrum of the work on visual scene analysis, though several pioneering works on the recognition or understanding of non-speech acoustic signals can been found in the literature (Oppenheim & Nawab, 1992; Nakatani et al., 1994; Ellis, 1994; Brown & Cooke, 1994). Here we consider two issues: flexibility of processing and the hierarchical structure of an auditory scene.

The flexibility of existing systems is rather limited when compared with human auditory abilities. For example, automatic music transcription systems that can deal with music played by a multi-instrument ensemble have not yet been developed, although several studies have been conducted (Roads, 1985; Mont-Reynaud, 1985; Chafe et al., 1985).

Recent progress in physiological and psychological acoustics has provided a wealth of information on the flexibility of auditory functions in humans. Of particular importance is that the property of information integration in the human auditory system has been elucidated, as demonstrated by the "auditory restoration" phenomenon (Handel, 1989). To achieve flexibility, mechanical systems must have this property because auditory scene analysis is an inverse problem in general formalization and cannot be properly solved without memory of sounds or models of the external world, as well as given sensory data.

An information integration model for sound understanding has already been devised using the blackboard architecture (Oppenheim & Nawab, 1992; Lesser et al., 1993). However, quantitative and theoretical treatments are still needed.

In addressing the structure issue, we introduce the concept of *perceptual sounds* as the hierarchical and symbolic representations of acoustic entities. The auditory stream (Bregman, 1990) is a familiar concept in auditory scene analysis: an auditory stream can be thought of as a cluster of acoustic energy formed in our auditory processes.

A perceptual sound, on the other hand, is a symbol that corresponds to an acoustic (or auditory) entity. In addition, an inherent property of perceptual sounds is its hierarchical structure, as will be shown in the following sections. Thus, auditory scene analysis will also be referred to more precisely as perceptual sound organization in this paper.

Against this background, we propose a novel process model of hierarchical perceptual sound organization with a quantitative information integration mechanism. The model is based on probability theory and characterized by its autonomous behavior and theoretically proved stability.

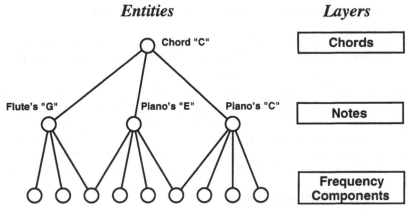

FIG. 9.1. Example of a snapshot of perceptual sounds.

9.2 PROBLEMS

9.2.1 Perceptual Sound Organization

One problem in perceptual sound organization is how to cluster acoustic energy to create the clusters that humans hear as one sound entity. Here it is important to note that humans recognize various sounds in a hierarchical structure in order to properly grasp and understand the external world. That is, a perceptual sound is structured in both spatial and temporal hierarchy. For example, when standing on a busy street, one sometimes hears all of the traffic noise as one entity, and at other times hears one specific car as one entity. If one directs attention to a specific car, the engine noise of that car and the sound caused by friction between the tires and pavement might be heard as two separate entities.

Figure 9.1 shows an example of a snapshot of perceptual sounds for music. For simplicity only the spatial structure is shown; the figure does not show temporal clusters of perceptual sounds like melodies or chord progression.

The problem of perceptual sound organization can be decomposed into three subproblems:

(1) Extraction of frequency components with an acoustic energy representation.

(2) Clustering of frequency components into perceptual sounds.

(3) Recognition of relationships among clustered perceptual sounds and building a hierarchical and symbolic representation of acoustic entities.

We consider the problem as extraction of *symbolic* representation from flat energy data, whereas some other approaches to auditory scene analysis relied on (i.e. evaluated their systems in terms of) restoration of target sound *signals* (Nakatani et al., 1994; Brown & Cooke, 1992). In the computer vision field, the scene analysis problem is considered one of extracting symbolic representations from bitmap images and is clearly distinguished from the image restoration problem, which involves the recovery of target images from noise or intrusions.

9.2.2 Music Scene Analysis

We chose music as an example of an applicable domain of perceptual sound organization. Our terminology is defined in Table 9.1. The term *music scene analysis* is used in the sense of perceptual sound organization in music. Specifically, it refers to the recognition of the frequency components, notes, chords, and rhythm of performed music.

Table 9.1
Summary of our terminology

Words	Meanings
perceptual sound	A symbol which represents an arbitrary acoustic event in the external world.
perceptual sound separation	Extraction of perceptual sounds from incoming sound signals
perceptual sound organization (auditory scene analysis)	Construction of an internal model of external acoustic events in a spatial and temporal structure with separation and restoration of perceptual sounds
music scene analysis	Auditory scene analysis for music sound signals

In the following sections, we first introduce general configuration of the music scene analysis system. We then focus the discussion on the hierarchical integration of multiple sources of information, which is an crucial problem in perceptual sound organization. Then behavior of the system and results of the performance evaluation are presented, followed by discussions and conclusions.

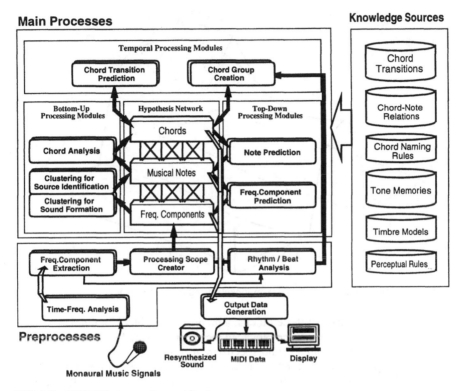

FIG. 9.2. OPTIMA processing architecture.

9.3 SYSTEM DESCRIPTION

Our processing architecture is called OPTIMA (Organized Processing toward Intelligent Music Scene Analysis) and it is illustrated in Figure 9.2. Input is assumed to be monaural music signals. The model creates hypotheses of frequency components, musical notes, chords, and rhythm. As a consequence of the probability propagation of hypotheses, the optimal (here used to mean "maximum likelihood") set of hypotheses is obtained and outputted as a score-like display, MIDI (Musical Instrument Digital Interface) data, or resynthesized source-separated sound signals.

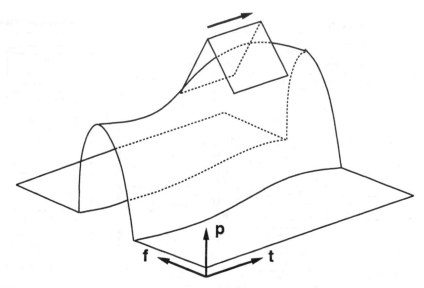

FIG. 9.3. Extraction of frequency components using pinching planes.

OPTIMA consists of three blocks: a preprocessing block, a main processing block, and a knowledge-source block. In the preprocessing block, first frequency analysis is performed and a sound spectrogram is obtained.

With this acoustic energy representation, frequency components are extracted. This process solves the first sub-problem mentioned in the previous section. Since it is difficult to achieve practical accuracy with a simple threshold method, we developed a method called the pinching plane to facilitate peak picking and tracking. As illustrated in Figure 9.3, two planes pinch a spectral peak, which allows us to find the temporal continuity of the spectral peaks. The planes are the regression planes of the peak, and are calculated by a least-squares fitting.

For complicated spectrum patterns, it is difficult to recognize onset and offset time solely based on bottom-up information. Thus the system creates several terminal point candidates for each extracted component. Using Rosenthal's rhythm recognition method (Rosenthal, 1992) and Desain's quantization method (Desain & Honing, 1989), rhythm information is extracted for precise extraction of frequency components and the recognition of onset/offset time. Based on the integration of the beat probabilities and termination probabilities of terminal point candidates, the candidates' status is fixed at either continuous or terminated and *processing scopes* are formed. A processing scope is a group of frequency components whose onset times are close: Our experiments on human auditory characteristics have clarified that if the onset asynchrony of two frequency components is greater than a certain threshold, the two components cannot form a *note* (the value of the threshold is typically 80 ms, though the value differs with the frequencies or onset gradients of the components). The processing scope is used as a basic time clock for the main processes of OPTIMA, as will be shown later.

When each processing scope is created in the preprocessing block, it is passed to the main processing block as shown in Figure 9.2. The main block has a hypothesis network with three layers corresponding to three levels of abstraction: (1) frequency components, (2) musical notes and (3) chords. Each layer encodes multiple hypotheses. That is, OPTIMA holds an internal model of the external acoustic entities as a probability distribution in the hierarchical hypothesis space.

Multiple processing modules are arranged around the hypothesis network. The modules are categorized into three blocks: (a) bottom-up processing modules to transfer information from a lower to a higher level, (b) top-down processing modules to transfer information from a higher to a lower level, and (c) temporal processing modules to transfer information along the time axis. The processing modules consult knowledge sources if necessary. The following sections discuss the information integration in the hypothesis network and behavior of each processing module.

9.4 INFORMATION INTEGRATION IN THE HYPOTHESIS NETWORK

For information integration in the hypothesis network, we require a method to propagate the impact of new information through the network.

We employ Pearl's Bayesian network method (Pearl, 1986), which can fuse and propagate new information represented by probabilities through the network using two separate links (λ-link and π-link) if the network is a singly connected (e.g. a tree-structured) graph.

Figure 9.4 shows our application of the hypothesis network. The three layers, or levels, mentioned in the previous section are denoted the C (component)-level, N (note)-level, and S (chord)-level. The link between the C-level and the N-level nodes is called the S (single)-link, which corresponds to one processing scope. The link between the S and N-levels becomes the M (multiple)-link, as a consequence of temporal integration: multiple notes along the time axis may form a single chord. The S-level nodes are connected timewise by the T (temporal)-link, which encodes chord progressions.

To discuss the information integration scheme, let us assume we wish to find the belief (BEL) induced on Node A in Figure 9.4 for example. Letting D_A^- stand for the data contained in the tree rooted at A and D_A^+ for the data contained in the rest of the network, we have

$$BEL(A) = P(A \mid D_A^+, D_A^-) \qquad \text{EQ. 9.1}$$

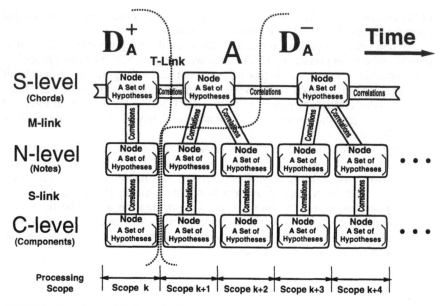

FIG. 9.4. Topology of the hypothesis network.

where A is a probability vector: $A = (a_1, a_2, ..., a_M)$. Using Bayes' theorem and assuming independence of hypotheses

$$P(D_A^+, D_A^- | a_j) P(D_A^- | a_j),$$ EQ. 9. 2

we then have

$$P(A | D_A^+, D_A^-) = aP(D_A^- | A) P(A | D_A^+),$$ EQ. 9. 3

where α is a normalization constant.

Substituting $\lambda(A) = P(D_A^- | A)$ and $\pi(A) = P(A | D_A^+)$ into Equation 9.3, it can be written as

$$BEL(A) = \alpha\lambda(A)\pi(A).$$ EQ. 9. 4

Given conditional probabilities P (Child|Parent) between any two adjacent nodes, $\lambda(A)$ can be derived from λ(Children of A) and $\pi(A)$ from π(Parent of A) (Pearl, 1986). This derivation is considered as the propagation of diagnostic (λ) or causal (π) support to A.

The minimum set of processing modules required in each node of the network is shown in Figure 9.5. B-Holder holds the belief (BEL) and passes new

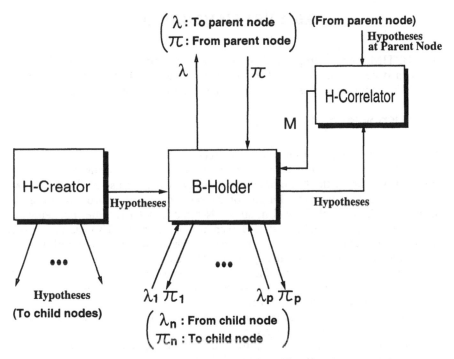

FIG. 9.5. Processing modules for each node of hypothesis network.

information as λ and π messages to adjacent B-Holders. In the OPTIMA model, B-Holders are embedded in the hypothesis network so there are not shown in Figure 9.2. H-Creator creates the hypotheses with initial probabilities. H-Correlator is for evaluating conditional probabilities $P(\text{Node}_1|\text{Node}_2)$, where Node_2 is a parent of Node_1, which are required in the information propagation process.

The minimum set of processing modules required in each node of the network is shown in Figure 9.5. B-Holder holds the belief (*BEL*) and passes new information as λ and π messages to adjacent B-Holders. In the OPTIMA model, B-Holders are embedded in the hypothesis network so there are not shown in Figure 9.2. H-Creator creates the hypotheses with initial probabilities. H-Correlator is for evaluating conditional probabilities $P(\text{Node}_1|\text{Node}_2)$, where Node_2 is a parent of Node_1, which are required in the information propagation process.

9.5 SYSTEM BEHAVIOR

The OPTIMA processing architecture has been implemented a music scene analysis system. This section discusses the configuration of knowledge sources and the behavior of processing modules in the main processing block.

9.5.1 Knowledge Sources

OPTIMA has six types of knowledge sources.

The chord transition dictionary holds statistical information on chord progressions, under the N-gram assumption (typically we use N=3); that is, we currently assume that the length of Markov chain of chords is three for simplicity. Since each S-level node has N-gram hypotheses, the independence condition in Equation 9.2 is satisfied even in S-level nodes. This dictionary was constructed based on a statistical analysis of 206 traditional songs (all western tonal music) that are popular in Japan and other countries.

In the chord-note relation database, the probabilities that certain notes will be played in a given chord are stored. This information was obtained by statistical analysis of the 2365 chords.

The chord-naming rules from music theory are used to name chords when hypotheses of played notes are given.

The tone memory is a repository of the frequency-component data of a single note played by various musical instruments (Figure 9.6). Currently it contains notes played by five instruments (clarinet, flute, piano, trumpet, and violin) at different expressions (forte, medium, piano), frequency ranges, and durations. We recorded those samples at a music studio.

The timbre models are formed in the feature space of the timbre. We first selected 41 parameters for musical timbre, such as the onset gradient of the frequency components and deviations in frequency modulations, and then reduced the number of parameters by the principal component analysis. With the proportion value of 95%, we have an eleven-dimensional feature space where at least the timbres of the foregoing five instruments are completely separated from each other. Assuming that one category of timbre has a normal distribution in the timbre space, we use $\bar{x}_j$, the averaged value of j-th parameter $(j = 1, 2, ..., m)$, and the variance-covariance matrix V as timbre model parameters for timbre category A.

Using Mahalanobis' distance D_i^2, the probability P that the i-th note belongs to category A can be calculated as

$$P = \frac{1}{(2\pi)^{m/2}\sqrt{|S|}} \exp\left\{-\frac{1}{2}D_i^2\right\}$$

EQ. 9. 5

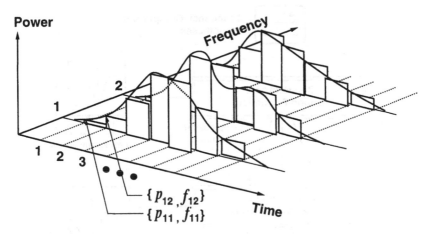

FIG. 9.6. Tone memory.

where $S = V^{-1}$ and

$$D_i^2 = \sum_j \sum_k (x_{ij} - \bar{x}_j) S_{jk} (x_{ik} - \bar{x}_k).$$
EQ. 9. 6

Finally, the **perceptual rules** describe the human auditory characteristics of sound separation (Bregman, 1990). Currently, the harmonicity rules and the onset timing rules are employed (Kashino & Tanaka, 1993) based on psychoacoustic experiments.

9.5.2 Bottom-Up Processing Modules

There are two bottom-up processing modules in OPTIMA: **NHC** (Note Hypothesis Creator) and **CHC** (Chord Hypothesis Creator). **NHC** is an H-Creator for the note layer and performs clustering for sound formation and clustering for source identification to create note hypotheses (Figure 9.7). It uses the perceptual rules for the former, and the timbre models for the discrimination analysis of timbres to identify the sound source of each note. **CHC** is an H-Creator for the chord layer and creates chord hypotheses when note hypotheses are given. It refers to the chord-naming rules in the knowledge sources.

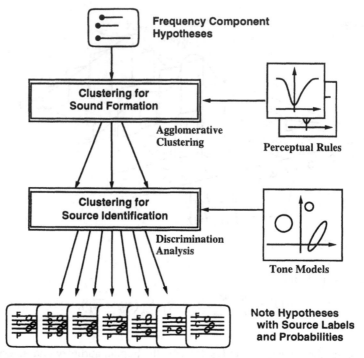

FIG. 9.7. Note hypothesis creator.

9.5.3 Top-Down Processing Modules

FCP (Frequency Component Predictor) and **NP** (Note Predictor) are the top-down processing modules. **FCP** is an H-Correlator between the note layer and the frequency component layer and evaluates conditional probabilities between hypotheses of the two layers by consulting tone memories (Figure 9.8). **NP** is an H-Correlator between the chord layer and the note layer and provides a matrix of conditional probabilities between those two layers (Figure 9.9). **NP** uses the stored chord-note relations knowledge.

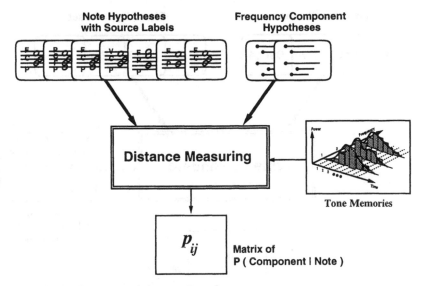

FIG. 9.8. Frequency component predictor.

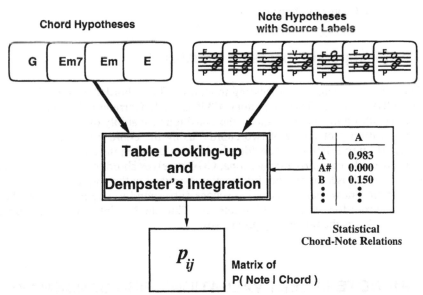

FIG. 9.9. Note predictor.

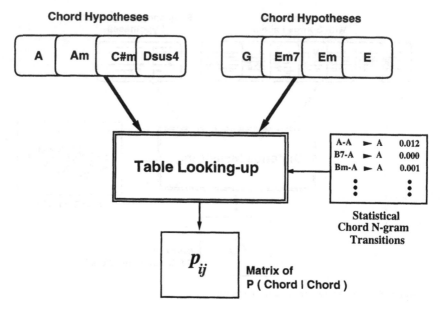

FIG. 9. 10. Chord transition predictor.

9.5.4 Temporal Processing Modules

There are also temporal processing modules: **CTP** (Chord Transition Predictor) and **CGC** (Chord Group Creator). **CTP** is an H-Correlator between the two adjacent chord layers. It estimates the transition probability of two N-grams (not the transition probability of two chords) using the chord transition knowledge source (Figure 9.10). **CGC** decides the M-Link between the chord layers and the note layers. In each *processing scope*, **CGC** receives chord hypotheses and note hypotheses. Based on rhythm information extracted in the preprocessing stage, it tries to find how many successive scopes correspond to one node in the chord layer in order to create M-Link instances. Thus the M-Link structure is formed dynamically as the processing progresses.

9.6 NOTE-LEVEL EVALUATION USING BENCHMARK TEST SIGNALS

We performed a series of tests to evaluate the system: frequency component level tests, note level tests, chord level tests, and tests using sample songs. In

FIG. 9. 11. Example of note patterns for note-level benchmark tests.

this section, we will concentrate on the note-level evaluation and present part of the results of the benchmark tests.

9.6.1 Test Patterns

In the note-level benchmark tests, we used simultaneous note patterns like the one shown in Figure 9.11. In the patterns, a given number of simultaneous notes (typically two or three) were continuously repeated by a MIDI sampler using digitized acoustic signals (16 bit, 44.1 kHz) of natural musical instruments (clarinet, flute, piano, trumpet, and violin). Note patterns were randomly composed by a computer, with one of the following constraints:

(1) Class 1 pattern: The interval between at least two of simultaneous notes is always an octave. That is, in the case of a two-simultaneous-note pattern, the fundamental frequencies of the two notes are always in harmonic relation.

(2) Class 2 pattern: The interval between at least two simultaneous notes is always a perfect fifth. That is, in the case of a two-simultaneous-note pattern, the second (fourth, sixth, ...) harmonic of one note and the third (sixth, ninth, ...) harmonic of another note always overlap.

(3) Class 3 pattern: Note patterns that do not belong to class 1 or class 2.

9.6.2 Parameters for Evaluation

The note recognition index R was defined as

$$R = 100 \cdot \left(\frac{right - wrong}{total} \cdot \frac{1}{2} + \frac{1}{2} \right) [\%], \qquad \text{EQ. 9.7}$$

where *right* is the number of correctly identified and correctly source-separated notes, *wrong* is the number of spuriously recognized (surplus) notes and incorrectly identified notes, and *total* is the number of notes in the input. Since it

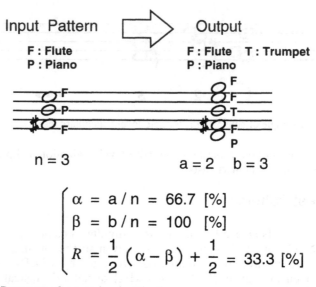

FIG. 9. 12. Parameters for note-level evaluation.

is sometimes difficult to distinguish surplus notes from incorrectly identified notes, both are included together in *wrong*. Scale factor 1/2 is for normalizing R: number of input notes, R becomes 0 (%) if all the notes are incorrectly identified and 100 (%) if all the notes are correctly identified by this normalization.

In addition, we also use the retrieval index α and the precision index β for the note-level evaluation:

$$\beta = \frac{b}{n},$$ EQ. 9. 8

$$\alpha = \frac{a}{n},$$ EQ. 9. 9

where n is the number of total notes in the input, a is the number of correctly identified and correctly source-separated notes in the output, and b is the number of the other notes in the output (Figure 9.12). R can be written using α and β as

$$R = \frac{1}{2}(\alpha - \beta) + \frac{1}{2}.$$ EQ. 9. 10

9.6.3 Results

The tests were performed with perceptual sound organization without any information integration (case 1), with information integration at the N-level only (case 2), and with total (i.e. N-level and S-level) information integration (case 3). In the first case, the best note hypothesis produced by the bottom-up processing (**NHC**) is just viewed as the answer by the system, while in the other two cases the tone memory information given by **FCP** is integrated.

Experimental results for the N-level evaluation is displayed in Figures 9.13 and 9.14 (class 1), Figures 9.15 and 9.16 (class 2), and Figures 9.17 and 9.18 (class 3). The α - β plots show the results for cases 1 and 2.

These results clearly indicate that integration of tone memory information significantly improves note recognition accuracy. Especially in the class 1 patterns, where most frequency components are overlapped, the α values before information integration was quite low. However, **NHC** creates octave hypotheses when there are only low-probability hypotheses, indicating the note-level information integration works well and consequently produces improved α values.

It is natural that chord-level integration did not affect the results, because the note patterns used in the tests were randomly (under the class constraint) composed and had nothing to do with the stored chord transition knowledge.

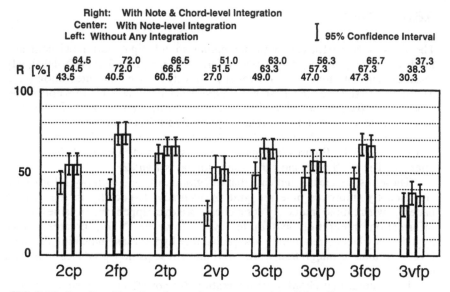

FIG. 9.13. Results of benchmark tests for note recognition (class 1).

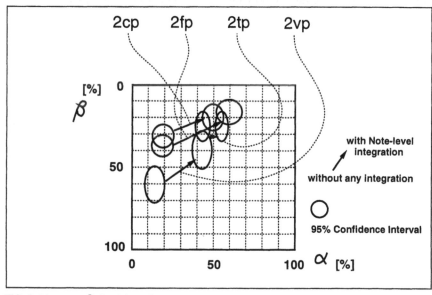

FIG. 9.14. $\alpha - \beta$ plot (class 1).

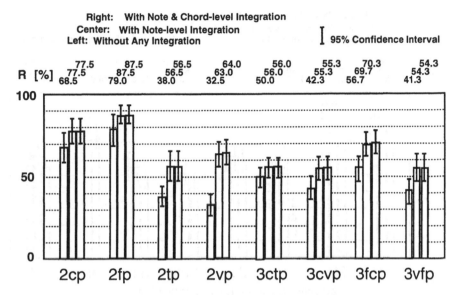

FIG. 9.15. Results of benchmark tests for note recognition (class 2).

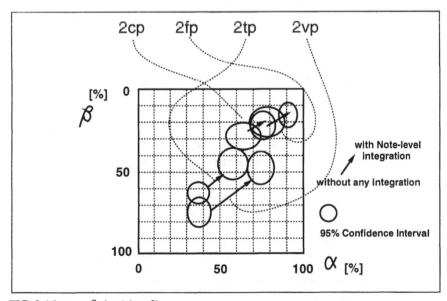

FIG. 9.16. α — β plot (class 2).

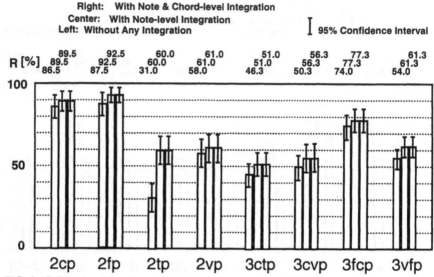

FIG. 9.17. Results of benchmark tests for note recognition (class 3).

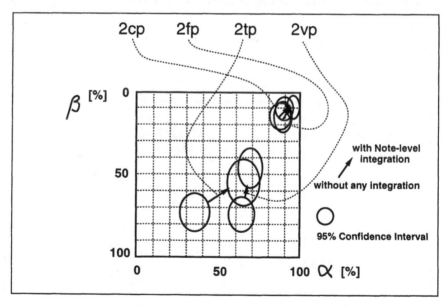

FIG. 9.18. $\alpha - \beta$ plot (class 3).

9.7 CONCLUSION

We first discussed the problem of perceptual sound organization in terms of constructing a valid symbolic hierarchical representation from incoming acoustic energy. Because we believe that a critical technical issue in this problem is the quantitative integration of multiple sources of information, we introduced an information integration scheme based on the Bayesian probability network and applied it to a music scene analysis system.

The Bayesian probability network enables stable information integration without any global control knowledge. The experimental results show that the integration of tone memory information significantly improves the recognition accuracy for perceptual sounds compared to conventional bottom-up based processing. In the situation when most frequency components in multiple notes overlap (e.g. class 1 and class 2), it was found that tone memory information is especially important.

This work focused on note-level evaluation. However, we have also found that the chord-level information is very effective too. See our other paper (Kashino et al., 1995).

One of the limits of our method lies in the topological constraint of the probability network: the probability propagation scheme described in this chapter cannot be applied to a multiply-connected graph (e.g., a mesh-structured graph). In our music scene analysis system, the most common error at the note level is misidentification of instruments, and the second most common error is in overtone pitch.

If we could introduce note-level transition information to the system, as well as the chord-level transition information, the recognition accuracy would be further improved; For example, information like "it is not usual that one note from a trumpet suddenly cuts into a flute and piano duet," or "this note is unusually high in pitch judging from this melody stream" would be useful. Thus, developing an information integration method applicable to a multiply-connected graph is the next step.

REFERENCES

Bregman, A. S. (1990). *Auditory scene analysis.* MIT Press.

Brown, G. J. & Cooke, M. (1992). A computational model of auditory scene analysis. In *Proceedings of International Conference on Spoken Language Processing*, pp. 523-526.

Brown, G. J. & Cooke, M. (1994). Perceptual grouping of musical sounds: A computational model. *Journal of New Music Research*, 23 (1), 107-132.

Chafe, C., Kashima, J., Mont-Reynaud, B., & Smith, J. (1985). Techniques for note identification in polyphonic music. In *Proceedings of the 1985 International Computer Music Conference*, pp. 399-405.

Desain, P. & Honing, H. (1989). Quantization of musical time: A connectionist approach. *Computer Music Journal*, 13 (3), 56-66.

Ellis, D. P. W. (1994). A computer implementation of psychoacoustic grouping rules. In *Proceedings of 12th International Conference on Pattern Recognition.* 3, 108-112.

Handel, S. (1989). *Listening.* MIT Press.

Kashino, K. & Tanaka, H. (1993). A sound source separation system with the ability of automatic tone modeling. In *Proceedings of the 1993 International Computer Music Conference*, pp. 248-255.

Kashino, K., Nakadai, K., Kinoshita, T. & Tanaka, H. (1995). Organization of hierarchical perceptual sounds : Music scene analysis with autonomous processing modules and a quantitative information integration mechanism. In *Proceedings of the 1995 International Joint Conference on Artificial Intelligence*, 1, 165-172.

Lesser, V., Nawab, S. H., Gallastegi, I. & Klassner, F. (1993). IPUS: An architecture for integrated signal processing and signal interpretation in complex environments. In *Proceedings of the 11th National Conference on Artificial Intelligence*, pp. 249-255.

Mont-Reynaud, B. (1985). Problem-solving strategies in a music transcription system. In *Proceedings of the 1985 International Joint Conference on Artificial Intelligence*, pp. 916-918.

Nakatani, T., Okuno, H. G., & Kawabata, T. (1994). Auditory stream segregation in auditory scene analysis with a multi-agent system. In

Proceedings of the 12th National Conference on Artificial Intelligence, pp.100-107.

Oppenheim, A. V. & Nawab, S. H. (Eds.). (1992*). Symbolic and Knowledge-Based Signal Processing.* Englewood Cliffs, NJ: Prentice Hall.

Pearl, J. (1986). Fusion, propagation, and structuring in belief networks. *Artificial Intelligence*, 29 (3), 241-288.

Roads, C. (1985). Research in music and artificial intelligence. *ACM Computing Surveys*, 17 (2), 163-190.

Rosenthal, D. (1992). Machine rhythm: computer emulation of human rhythm perception. Ph.D. Thesis, Department of Computer Science, Massachusetts Institute of Technology.

10

Context-Sensitive Selection of Competing Auditory Organizations: A Blackboard Model

Darryl Godsmark and Guy J. Brown
University of Sheffield

It was proposed (Bregman, 1990) that auditory organization is performed by a wide range of grouping principles. One problem facing any computational model incorporating a number of these principles is that of resolving the conflict arising when several principles suggest mutually exclusive organizations. This chapter presents a principled computational framework within which conflicts are resolved by a context-sensitive evidence accumulation mechanism. The mechanism exhibits several properties consistent with human auditory organization. Additionally, as grouping principles operate in complete independence, the framework allows any number of grouping principles to be incorporated.

10.1 INTRODUCTION

Bregman's theory of auditory scene analysis (Bregman, 1990) posits that auditory organization is conceptually a two–stage process. In the first stage, early auditory processes decompose the sound wave into a number of sensory elements. Subsequently, those elements likely to have originated from a common environmental source are grouped in a perceptual representation termed a stream. From as early as 1950 (Miller & Heise, 1950) it was suggested that this organizational process could be performed by grouping principles similar to those proposed for visual organization by the Gestalt psychologists (Koffka, 1936).

Subsequent psychoacoustic investigations (Bregman, 1990) suggested a wide variety of potential auditory grouping principles, including temporal and

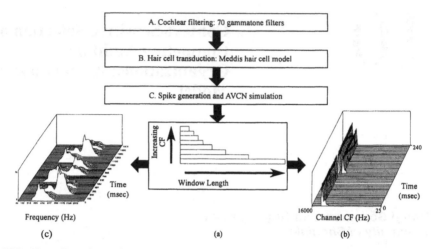

FIG. 10.1. Overview of the stages of the model simulating early auditory processes. Period (panel b) and firing rate (panel c) representations are calculated using variable width windows (panel a). The stimulus is a short section of the Bregman and Tougas [1989] stimulus shown in Fig. 10.3(a).

frequency proximity, common onsets and offsets, harmonicity, pitch continuity, coherent amplitude modulation and continuity of timbre.

Any computational model of auditory organization which intends to employ this plethora of grouping principles must incorporate a strategy for integrating the evidence provided by individual principles. More importantly, this strategy must also be capable of resolving the conflict arising when two or more principles suggest mutually exclusive organizations.

The remainder of this chapter presents a computational implementation of a grouping strategy which addresses both of these problems in a manner compatible with known properties of human auditory organization. The model consists of two main processing stages; the first stage simulates the auditory periphery, and certain early neural processes, wherein particular features of the auditory signal are made explicit. Subsequently, these features are used to drive the grouping mechanism, which forms the second major stage of processing. The chapter is structured to mirror the order of processing within the model.

10.2 FEATURE IDENTIFICATION

Thus far, the stimuli employed as input to the model have been composed entirely of sequences of pure tones. Consequently, the auditory front-end and feature identification processes were deliberately designed to exploit the simple properties of pure tones, and are highly functional. These processes are summarized in Figure 10.1.

10.2.1 The Auditory Periphery

The frequency-selective properties of the basilar membrane are simulated by a bank of 70 phase-corrected gammatone filters (Patterson, Holdsworth, Nimmo–Smith & Rice, 1988). The center frequencies of these filters are linearly distributed on the ERB-rate scale (Glasberg & Moore, 1990) between 50Hz and 2500Hz. For reasons of computational efficiency, the filterbank is restricted to the range of frequencies employed by the input stimulus.

Pressure gains introduced by the outer and middle ear are combined into a single transfer function (based on an equation from ISO standard 226: Acoustics–normal equal loudness level contours), which is used to directly modify the gain of each gammatone filter.

10.2.2 Auditory Nerve Spike Generation

The conversion of basilar membrane motion to neural activity by the inner hair cells (Pickles, 1988) is simulated by a two–stage process. Initially, the Meddis hair cell model (Meddis, 1988) is applied to each filterbank channel, producing the probability of a nerve spike occurring at each instant in time. In a review of hair-cell models (Hewitt & Meddis, 1991), the Meddis model was found to most closely simulate a number of physiological findings, including both phase-locking and adaptation.

A particularly abstract approach to spike generation is employed which simulates activity in the auditory nerve and certain processes in the anteroventral cochlear nucleus (AVCN). In keeping with physiological studies (see Handel (1989)), it is assumed that the output from the hair cell model excites a population of auditory nerve fibers (in this case 20). Furthermore, it was suggested (Carney, 1992) that certain cells in the AVCN may perform some kind of nerve spike coincidence detection. A number of auditory nerve fibers converge on a single cell, which will itself fire only when spikes appear simultaneously in a given number of these input fibers. Carney argued that this could be the basis of a signal-to-noise ratio enhancement mechanism, as spectral peaks will be highly correlated across fibers (and thus cause the AVCN cells to fire) whereas noise and spontaneous activity will be much less correlated, and thus be filtered.

This process is grossly approximated by assuming that spikes occur only in coincidence with peaks in the hair cell firing probabilities, and the number of responding fibers is directly proportional to the magnitude of this peak. Finally, much of the spontaneous activity in the population response is removed by applying a threshold. All future processing is performed on the response of the population as a whole rather than individual fibers.

10.2.3 Firing Rate and Periodicity Analysis

As mentioned above, the stimuli employed to train and test the model have thus far been composed entirely of pure tones. Accordingly, there are only four salient features:

(1) Onsets

(2) Offsets

(3) Frequency

(4) Loudness

These features are extracted from two representations, based upon average firing rates and periodicity. These representations are discussed in detail below.

Periodicity Representation. In a given channel, periodicity analysis involves counting the time delay between each spike and every other spike within a given window. Following a scheme proposed by Ghitza (1986), for each channel there are a number of bins (currently 80) that represent time intervals between spikes, with each successive bin representing a larger time interval than its predecessor. Within the window, each time a particular interval between spikes is detected, the contents of the appropriate bin are incremented.

The periodicity analysis is performed over a window just wide enough to encompass one complete period of the lowest frequency to which a given filter will respond. Consequently, the width of the window varies for each channel, with the width diminishing with increasing center frequency (Figure 10.1(a)).

Finally, the periodicity analysis for each individual channel is combined into a single representation by summing equivalent bins across every channel (see Figure 10.1(b)), leading to a representation similar to the summary autocorrelogram proposed by Meddis and Hewitt (1991), but considerably more efficient to calculate. Furthermore, the representation employed here exhibits significantly more prominent peaks than are normally associated with the summary autocorrelogram representation.

Firing Rate Representation. The second representation employed by the model is a representation of average firing rates. Essentially, this is a running sum of how many fibers have fired in each channel within a given temporal window. Again, this sum is performed within a window just wide enough to include one period of the lowest frequency the channel can respond to. From this representation (see Figure 10.1(c)), onsets and offsets can be easily detected by noting that:

(1) The onset of a pure tone will be accompanied by a very sudden increase in the firing rate.

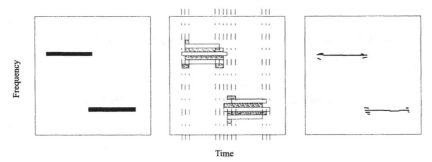

FIG. 10.2. An example of component formation, showing how tones (left panel) excite a range of filters, which are segmented into areas with stable properties (center panel). Finally, components with similar periodic responses are combined to form components with a specific frequency, onset and offset (right panel).

(2) The offset of a pure tone will be accompanied by a very sudden decrease in the firing rate.

In practice, for the stimuli considered here, because spontaneous activity has been filtered out onsets are indicated by the firing rate increasing from zero, and offsets are indicated by a return to a zero firing rate.

The firing rate representation could potentially be used to derive the loudness of a tone, but grouping by loudness has currently not been implemented.

10.2.4 Component Formation

On the basis of the above representations, each channel is segmented into a number of auditory components. Components are essentially a computational convenience which delineate areas of relatively stable properties. Components are formed by applying a set of simple rules, based upon the synchrony strands scheme proposed by Cooke (1993), to each channel (see also (Brown & Cooke, 1994)):

(1) If an onset is detected, start a new component and flag it as possessing a definite onset.

(2) If an offset is detected, terminate the component and flag it as possessing a definite offset.

(3) If a significant (determined by a threshold) change in frequency (periodic response) is detected, start a new component and label it with the new frequency.

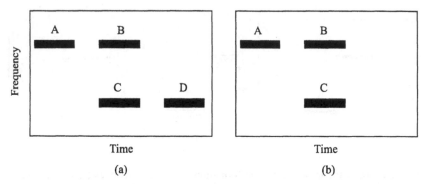

FIG. 10.3. The Bregman and Tougas (1989) stimulus with tone D present (a) and replaced with a silence (b).

   (4) If a region has a firing rate of zero, mark it as silence and ignore it in all
        future processing.

Additionally, when a component in one channel is started or terminated, new components are simultaneously started in all other channels. This is because changes in a channel may trigger a new organization, potentially capturing other (stable) channels. As components form the fundamental unit of grouping, a new component must be started in every channel to allow this. An example of component formation is shown in Figure 10.2.

   As noted by Cooke (1993), given that gammatone filters respond to a range of frequencies, and are spaced such that they overlap, a single tone will initiate activity in a number of adjacent filters. Much of this activity is therefore redundant, and can be removed by combining components which exhibit a similar periodic response (i.e., stimulate the same or adjacent bins).

10.3 AUDITORY ORGANIZATION

The grouping strategy proposed here is motivated by observations concerning human auditory grouping:

   (1) Organizations can be formed retroactively; that is, an organization does
        not immediately have to be imposed upon components.

   (2) Organization is context-sensitive; that is, it is not just the relationship
        between adjacent components that determines their organization – their
        relationship to surrounding components must also be considered.

Although there have been a number of experiments which demonstrate both retroactive organization (e.g., (Dannenbring & Bregman, 1978)) and context-sensitive organization (e.g., (Bregman & Rudnicky, 1975)), perhaps the most striking example of both was presented by Bregman and Tougas (1989).

They presented subjects with repetitive cycles of the two stimuli shown in Figure 10.3. Subjects were asked to "evaluate the clarity of the two high tones as a pair." If tone D is present (Figure 10.3(a)) then tones A and B are quite clearly perceived as a separate stream to the tones C and D. In contrast, if tone D is replaced with an equivalent duration of silence (Figure 10.3(b)) then the clarity of tones A and B as a pair is greatly reduced. Bregman and Tougas argue that this is due to tones B and C fusing to form a complex tone, so the sequence is perceived as a pure tone followed by a complex tone.

Thus the fusion of tones B and C is dependent not only on their relationship to each other, but also on the presence or absence of tone D, demonstrating the context-sensitive nature of auditory grouping. Furthermore, the organization of tones B and C cannot be finalized until the presence or absence of tone D has been established, demonstrating retroactive organization.

10.3.1 Grouping Strategy

The assertion that organization is both retroactive and context-sensitive suggests a possible strategy for resolving conflicts between grouping principles. Rather than making an immediate, arbitrary decision as to which organization will be imposed, all potential organizations are retained in the hope that the evolving context will provide enough additional evidence to resolve the conflict.

Organization Hypothesis Region. Obviously, the auditory system must wait a finite time before it enforces an organization. Thus, the model employs a sliding temporal window, which represents the organization hypothesis region (OHR). Essentially, alternative organizations are only retained for components within the OHR. Once a component leaves the OHR, an ultimate organization for that component must be imposed.

One problem with this approach is determining the duration of the OHR. However, it appears that the phenomenon of perceptual restoration (e.g., (Warren, 1984)) represents a further example of retroactive grouping. Here, a section of a sound obliterated by a noise burst may be perceptually reconstructed, on condition that the evidence available when the noise burst terminates suggests the sound was masked, as opposed to replaced, by the noise. This reconstruction must therefore be performed retroactively.

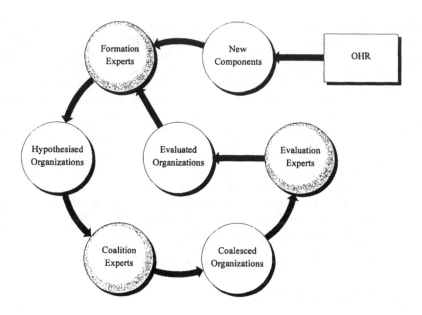

FIG. 10.4. Overview of the lower levels of the blackboard, showing the cyclic nature of organisation formation. Levels of the blackboard are shown as clear circles, and peers (groups of experts who operate at the same level of the blackboard) are represented by a single shaded circle.

Assuming that this is an example of the same retroactive mechanism discussed earlier, the maximum duration of noise through which perceptual restoration occurs provides an indication of the maximum duration of the OHR. Current estimates (Kluender & Jenison, 1992) place this value between 250 and 350 msec. In the current implementation, a window length of 250 msec was selected.

10.3.2 Organization Cycle

Essentially, organization formation is a cyclic process. As the OHR progresses through the auditory signal, hypotheses are formed concerning the organization of components entering the OHR. These hypotheses are scored on the basis of how closely they abide by various grouping principles, and form the basis of the next cycle of grouping.

In practice, this organizational cycle has been implemented as a blackboard architecture (see Engelmore & Morgan (1988) for reviews), and consists of four distinct phases; organization hypothesis formation, organization coalition, organization evaluation and organization selection. These phases can be seen in Figure 10.4.

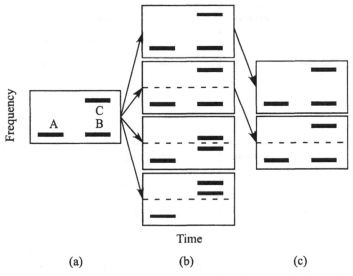

Time

(a)   (b)   (c)

FIG. 10.5. An example of hypothesis formation. Given the organization shown in (a), with tones B and C having just entered the OHR, there are four possible organizations (b). However, given the relationship of tones B and C to tone A, only two of these hypotheses are retained as plausible organizations. The broken lines indicate tones in different streams.

Organization Hypothesis Formation. When components enter the OHR, it must be decided whether they naturally extend an existing hypothesized organization, or if a new hypothesis must be created. This decision is performed by a collection of formation experts, one expert for each grouping principle. In complete isolation, each expert applies heuristics to determine plausible organizations according to the grouping principle represented by that expert.

The best way to understand this process is through an example; consider the situation shown in Figure 10.5(a). A single tone (A) has entered the OHR, producing only one possible organization. Some time later, two more tones (B and C) simultaneously enter the OHR and are considered by the frequency proximity formation expert. In total, there are 4 possible organizations, shown in Figure 10.5(b).

To the frequency proximity expert, tones A and B – being the same frequency – represent an ideal relationship; hence any organization in which these two tones are segregated is deemed implausible and is rejected. In contrast, in terms of frequency tone C is some distance away from tone A, hence it is equally plausible to suggest that tones A and C group with each other, or are placed in a separate stream. Thus a frequency proximity formation expert would create two organization hypotheses (Figure 10.5(c)).

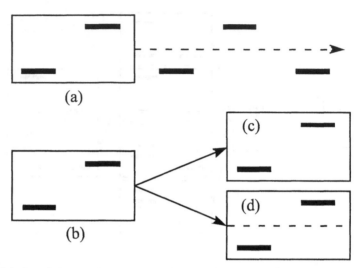

FIG. 10.6. An example of hypothesis evaluation. The OHR slides over the stimulus (a), producing hypotheses (b,c and d) for evaluation. See text for further details.

As this process proceeds, other experts (for example, temporal proximity or onset asynchrony) may suggest alternative organizations, leading to the creation of independent sets of organization hypotheses.

Organization Coalition.   During the hypothesis formation stage, it is probable that a range of experts will independently suggest identical organizations, albeit for different reasons. The organization coalition expert merges the sets of organization hypotheses, and in the process removes these redundant repetitions, forming a single set of unique organization hypotheses.

Organization Evaluation.   Each organization hypothesis is evaluated by a range of evaluation experts, one for each grouping principle. The process of evaluation is best explained by an example; consider the stimulus shown in Figure 10.6(a) being processed by temporal and frequency proximity experts only. The organization hypotheses region slides from left to right over the stimulus, and hypotheses are formed about the enclosed components.

When the first component enters the OHR, there is only one possible organization (Figure 10.6(b)). As the second component enters the OHR, there are two possible organizations; either the two components are placed in a single stream (Figure 10.6(c)) or in different streams (Figure 10.6(d)). In both cases, the organizations are independently evaluated by the temporal and frequency proximity experts, who give the organization a score representing the plausibility of that organization.

This score is directly dependent upon how closely the relationship between the newly added component(s) and the immediately preceding component(s) in each stream abides by the expert's principle. So, for example, the temporal

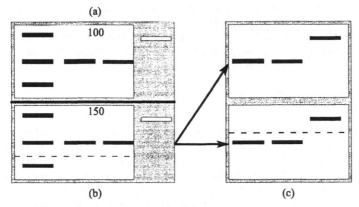

FIG. 10.7. As the three tones on the left in panels (a) and (b) leave the OHR, their organisation is determined by the hypothesis with the highest plausibility index (in this case, panel (b)). When a new tone (shown in white) enters the OHR, hypotheses concerning its organization (c) will be grown from the hypothesis shown in panel (b).

proximity expert would give the organization shown in Figure 10.6(c) a positive score; the two components are temporally proximate, so grouping them in the same stream is plausible. In contrast, the same expert would give the organization in Figure 10.6(d) a negative score, as segregating two temporally proximate components abuses the temporal proximity principle. The scores from each expert are then combined, producing an evaluation of the plausibility of the organization according to the population of experts, termed the plausibility index.

As the next component enters the OHR, the formation experts produce a range of potential organizations, using the two current organizations as seeds. A plausibility index will be calculated as described above for each new organization. Additionally, each new organization inherits the plausibility index of its immediate ancestor. The two indices are combined to give the overall plausibility index of the new organization. These organizations now form the seed for the next generation of organization hypotheses.

Organization Selection. When a component exits the OHR, the organization hypothesis with the highest plausibility index determines the organization that will be imposed upon that component, and all hypotheses that oppose that organization are terminated.

For example, consider the organization hypotheses shown in Figures 10.7(a) and 10.7(b). If the three simultaneous components exit the OHR, then their ultimate organization will be that shown in Figure 10.7(b), as it has the highest plausibility index. Given this, the organization shown in Figure 10.7(a) is no longer plausible and is deleted, so all future organization hypotheses will be grown from those shown in Figure 10.7(c).

As the plausibility index accumulates over time, this ensures that the organization of a component is strongly dependent upon the context in which it appears. Also, an organization which initially appears strong may be weakened - or alternatively initially weak organizations may be strengthened – by future components, providing a retroactive grouping mechanism.

10.3.3 Training Phase

One problem which must always be faced in a model of this nature is that of defining the evaluation metrics employed by the experts to produce the plausibility index. Previously, a number of models (e.g., (Kashino & Tanaka, 1992)) derived evaluation metrics directly from psychophysical data For example, Kashino and Tanaka performed an experiment to determine how easy it was to "hear out" a partial with varying degrees of onset asynchrony, and implemented an evaluation function directly from these findings.

However, this approach suffers from a number of problems, which are a direct consequence of the fact that subject responses are based on the stimulus being processed by the entire auditory system. The foremost problem is that auditory organization is context sensitive. As the stimulus is being processed by a context sensitive mechanism, the experiment is implicitly investigating the effect of onset asynchrony in a particular context. Furthermore, given that the auditory system is employing many grouping principles in addition to onset asynchrony, the experiment is measuring the combined effects of all the grouping principles. Indeed, it is difficult to conceive of an experimental paradigm which would allow the contribution of a single grouping principle to be isolated.

For the model proposed here, the evaluation metrics need to be applicable in any context; it is the grouping mechanism that introduces context effects. Similarly, the grouping mechanism combines the hypotheses of the various grouping principles, so each evaluation metric should represent an isolated grouping principle. Consequently, direct derivation of the evaluation metrics would appear to be impractical. Accordingly, a novel approach has been adopted, whereby experts are trained using a genetic algorithm paradigm (Goldberg, 1989).

Evaluation Metrics. Evaluation metrics are assumed to be curves, as shown in Figure 10.8, with two curves required for each grouping principle. For any two components, the curves express the relationship between the plausibility index (on the ordinate) and how closely the two components abide by the represented grouping principle (on the abscissa). If a hypothesis groups the components in the same stream, the solid curve determines the plausibility index; alternatively, the broken curve determines the plausibility index if the two components have been segregated. This segregation curve is simply the mirror image of the grouping curve.

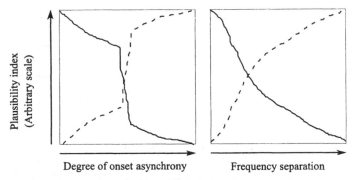

FIG. 10.8. Hypothetical evaluation metrics for onset asynchrony (left) and frequency proximity (right). The point where the solid and broken curves cross is the point at which fusion and segregation are equally plausible.

For each grouping principle, an independent population is maintained, which encodes the coordinates of a limited number of points on the curve. Blending functions (Harrington, 1983) are then employed to smoothly interpolate between these points, generating the full curve.

Although the genetic algorithm paradigm allows for the initial population to be random, in practice it was restricted to include only intuitively feasible curve pairs. Essentially, certain points on the curve can be approximately positioned in advance, and gross characteristics of the curve can be determined. For example, on the basis of frequency proximity any hypothesis that groups components with identical frequencies will merit a high plausibility index, so curves which don't meet this requirement can be immediately dismissed. Similarly, the plausibility index will decrease with increasing frequency separation, so curves exhibiting an increase can be rejected.

Training Cycle. The training cycle takes the form of a relatively complex incremental cycle, of which only a brief overview is presented here due to limited space. Initially, only two experts are implemented, representing temporal and frequency proximity. Using the curves represented by the initial populations, the model processes 15 seconds of a wide range of repetitive two-tone sequences, as employed by van Noorden (1977).

The output from the model is compared with published subject responses from a range of relevant psychoacoustic investigations (e.g., (van Noorden, 1977)), and the populations are evolved accordingly. The stimuli are presented to the model such that the temporal relationship between consecutive tones remains constant while various frequency separations are considered. The time delay between tones is then increased, and the same frequency separations are again considered, and so on. Once these two experts are trained, additional principles are incorporated incrementally.

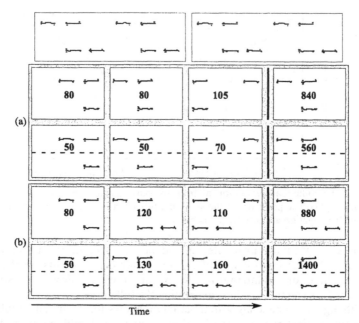

FIG. 10.9. Sample output from the model at successive instants in time, processing the Bregman and Tougas [1989] stimulus, with tone D omitted (panel a) and present (panel b). The top two boxes show the representation of the stimulus calculated by the model. In both panels, the two rows represent the two strongest hypotheses in which tones B and C are fused (top row) and segregated (bottom row). The first three boxes of each row represent hypotheses created during the first repetition of the stimulus, whereas the right-most boxes are the same hypotheses after 8 repetitions have been processed. The numbers are the overall plausibility index for each hypothesis.

10.4 RESULTS

The model is currently at an early stage of training, and as yet none of the experts are fully trained. However, despite the limited training, the model can demonstrate a qualitative match to a range of psychoacoustic data.

A sample of the models output at various stages of processing is shown in Figure 10.9. The stimulus being processed is the Bregman and Tougas stimulus discussed earlier (see Figure 10.3). As can be seen, with tone D absent (Figure 10.9(a)) the most favored organization fuses tones B and C, whereas with tone D present (Figure 10.9(b)), the organization in which tones B and C are segregated gradually becomes the predominant organization. This is consistent with the results reported by Bregman and Tougas (1989).

This example also demonstrates an interesting emergent property of the model; given a stable, repetitive stimuli, the plausibility index of the predominant organization continually increases. If the plausibility index is interpreted as the "strength" or clarity of the perceived organization, then the strength of the organization accumulates over time. Again, this is consistent with observations concerning auditory organization (Bregman, 1990).

10.5 CONCLUSIONS AND FUTURE WORK

This chapter presents a novel, context-sensitive grouping mechanism, providing a principled computational framework within which many perceptual grouping principles can be incorporated. At present, the model demonstrates a qualitative match to psychophysical data, but much additional training is required before a quantitative match can be obtained.

Following additional training, a second context-sensitive mechanism will be implemented to incorporate grouping on the basis of common pitch and timbre. As pitch and timbre are emergent properties of a fused set of components, these principles are assumed to operate at a higher level than the primitive principles discussed above.

The primary reason for implementing the grouping mechanism as a blackboard system is that future development will include top–down grouping. In the context of speech or music, the structured nature of the auditory signal allows predictions about future components to be formed, which could subsequently control an expectancy mechanism, strengthening predicted organizations.

10.6 ACKNOWLEDGEMENTS

Darryl Godsmark is supported by a grant from the BBSRC. Guy J. Brown is supported by SERC grant GR/H53174, EPSRC grant GR/K18962 and the Nuffield Foundation.

REFERENCES

Bregman, A. S. (1990). *Auditory Scene Analysis*. Cambridge, MA: MIT Press.

Bregman, A. S. & Rudnicky, A. I. (1975). Auditory segregation: stream or streams?. *Journal of Experimental Psychology: Human Perception and Performance*, 1(3): 263–267.

Bregman , A. S. & Tougas, Y. (1989). Propagation of constraints in auditory organization. *Perception and Psychophysics*, 46(4): 395–396.

Brown, G. J. & Cooke, M. (1994). Computational auditory scene analysis. *Computer Speech and Language*, 8: 297–336.

Carney, L. H. (1992). Modelling the sensitivity of cells in the anteroventral cochlear nucleus to spatiotemporal discharge patterns. *Philosophical Transactions of the Royal Society of London* (B), 336: 403-406.

Cooke, M. (1993). *Modelling auditory processing and organization.* Cambridge, U. K.: Cambridge University Press.

Dannenbring, G. L. & Bregman, A. S. (1978). Streaming vs. fusion of sinusoidal components of complex tones. *Perception and Psychophysics*, 24(4): 369–376.

Engelmore, R. & Morgan, T. (eds). (1988). *Blackboard systems*. Reading, MA: Addison-Wesley.

Ghitza, O. (1986). Auditory nerve representation as a front-end for speech recognition in a noisy environment. *Computer Speech and Language*, 1: 109–130.

Glasberg, B. & Moore, B. C. J. (1990). Derivation of auditory filter shapes from notched-noise data. *Hearing Research*, 47: 103–138.

Goldberg, D. E. (1989). *Genetic Algorithms in Search, Optimization and Machine Learning*. Reading: Addison-Wesley.

Handel, S. (1989). *Hearing: An Introduction: to the Perception of Auditory Events*. Cambridge, MA: MIT Press.

Harrington, S. (1983). *Computer Graphics: A Programming Approach.* New York: McGraw Hill Inc.

Hewitt, M. J. & Meddis, R. (1991). An evaluation of eight computer models of mammalian inner hair-cell function. *Journal of the Acoustical Society of America*, 90(2): 904–917.

Kashino, K. & Tanaka, H. (1992). A sound source separation system using spectral features integrated by the Dempster's law of combination. *Annual Report of the Engineering Research Institute*, Faculty of Engineering, University of Tokyo, 51: 67–72.

Kluender, K. R. & Jenison, R. L. (1992). Effects of glide slope, noise intensity and noise duration on the extrapolation of FM glides through noise. *Perception & Psychophysics*, 51(3): 231–238.

Koffka, K. (1936). *Principles of Gestalt Psychology.* New York: Harcourt and Brace.

Meddis, R. (1988). Simulation of auditory-neural transduction: further studies. *Journal of the Acoustical Society of America*, 83(3): 1056–1063.

Meddis, R. & Hewitt, M. J. (1991). Virtual pitch and phase sensitivity of a computer model of the auditory periphery. I: Pitch identification. *Journal of the Acoustical Society of America*, 89(6): 2866–2894.

Miller, G.A. & Heise, G. A. (1950). The trill threshold. *Journal of the Acoustical Society of America*, 22: 167–173.

Patterson, R. D., Holdsworth, J., Nimmo-Smith, I., & Rice, P. (1988). *Implementing a gammatone filterbank.* APU Report 2341.

Pickles. (1988). *An Introduction to the Physiology of Hearing* (second edition). London: Academic Press.

van Noorden, L. (1977). Minimum differences of level and frequency for perceptual fission of tone sequences ABAB. *Journal of the Acoustical Society of America*, 61(4): 1041-1045.

Warren, R. M. (1984). Perceptual restoration of obliterated sounds. *Psychological Bulletin*, 96(2): 371–383.

11

Musical Understanding at the Beat Level: Real-time Beat Tracking for Audio Signals

Masataka Goto and Yoichi Muraoka
Waseda University

This chapter presents the main issues and our solutions to the problem of understanding musical audio signals at the beat level, issues which are common to more general auditory scene analysis. Previous beat tracking systems were not able to work in realistic acoustic environments. We built a real-time beat tracking system that processes audio signals that contain sounds of various instruments. The main features of our solutions are:

(1) To handle ambiguous situations, our system manages multiple agents that maintain multiple hypotheses of beats.

(2) Our system makes a context-dependent decision by leveraging musical knowledge represented as drum patterns.

(3) All processes are performed based on how reliable detected events and hypotheses are, since it is impossible to handle realistic complex signals without mistakes.

(4) Frequency-analysis parameters are dynamically adjusted by interaction between low-level and high-level processing.

In our experiment using music on commercially distributed compact discs, our system correctly tracked beats in 40 out of 42 popular songs in which drums maintain the beat.

11.1 INTRODUCTION

Our goal is to build a system that can understand musical audio signals in a humanlike fashion. We believe that an important initial step is to build a system which, even in its preliminary implementation, can deal with realistic audio signals, such as ones sampled from commercially distributed compact discs. Therefore our approach is first to build such a robust system which can understand music at a low level, and then to upgrade it to understand music at a higher level.

Beat tracking is an appropriate initial step in computer understanding of Western music, because beats are fundamental to its perception. Even if a person cannot completely segregate and identify every sound component, he or she nevertheless can track musical beats and keep time to music by hand-clapping or foot-tapping. It is almost impossible to understand music without perceiving beats, because the beat is a fundamental unit of the temporal structure of music. We therefore first build a computational model of beat perception and then extend the model, just as a person recognizes higher-level musical events on the basis of beats.

Following these points of view, we build a beat tracking system, called *BTS*, which processes realistic audio signals and recognizes temporal positions of beats in real time. BTS processes monaural signals that contain sounds of various instruments and deals with popular music, particularly rock and pop music in which drums maintain the beat. Not only does BTS predict the temporal position of the next beat (quarter-note); it also determines whether the beat is strong or weak[1]. In other words, our system can track beats at the half-note level.

To track beats in audio signals, the main issues relevant to auditory scene analysis are:

(1) In the interpretation of audio signals, various ambiguous situations arise. Multiple interpretations of beats are possible at any given point, since there is not necessarily a single specific sound that directly indicates the beat position.

(2) Decisions in choosing the best interpretation are context-dependent. Musical knowledge is necessary to take a global view of the tracking process.

(3) It is almost impossible to detect all events in complex audio signals correctly and completely. Moreover any interpretation of detected events may include mistakes.

(4) The optimal set of frequency-analysis parameters depends on the input. It is desirable to adjust those parameters based on a kind of global context.

Our beat tracking system addresses the issues presented above. To handle ambiguous situations, BTS examines multiple hypotheses maintained by multiple agents that track beats according to different strategies. Each agent makes a context-dependent decision by matching pre-stored drum patterns with the currently detected drum pattern. BTS also estimates how reliable detected events and hypotheses are, since they may include both correct and incorrect interpretations. To adjust frequency-analysis parameters dynamically, BTS supports interaction between onset-time finders in the low-level frequency analysis and the higher-level agents that interpret these onset times and predict beats.

To perform this computationally intensive task in real time, BTS has been implemented on a parallel computer, the Fujitsu AP1000. In our experiment with 8 pre-stored drum patterns, BTS correctly tracked beats in 40 out of 42 popular songs sampled from compact discs. This result shows that our beat-tracking model based on multiple-agent architecture is robust enough to handle real-world audio signals.

11.2 ACOUSTIC BEAT-TRACKING ISSUES

The following are the main issues related to tracking beats in audio signals, and they are issues which are common to more general computational auditory frameworks that include speech, music, and other environmental sounds.

11.2.1 Ambiguity of Interpretation

In the interpretation of audio signals, various ambiguous situations arise. At any given point in the analysis, multiple interpretations may appear possible; only later information can determine the correct interpretation. In the case of beat tracking, the position of a beat depends on events that come after it. There are several ambiguous situations, such as ones where several events obtained by frequency analysis may correspond to a beat, and different interbeat intervals[2] seem to be plausible.

11.2.2 Context-Dependent Decision

Decisions in choosing the best interpretation are context-dependent. To decide which interpretation in an ambiguous situation is best, global understanding of the context or situation is desirable. A low-level analysis, such as frequency analysis, cannot by itself provide enough information on this global context. Only higher-level processing using domain knowledge makes it possible to make an appropriate decision. In the case of beat tracking, musical knowledge is needed to determine whether a beat is strong or weak and which note-value it corresponds to.

11.2.3 Imprecision in Event Detection

It is almost impossible to detect all events in complex audio signals correctly. In frequency analysis, detected events will generally include both correct and incorrect interpretations. A system dealing with realistic audio should have the ability to decide which events are reliable and useful. Moreover, when the system interprets those events, it is necessary to consider how reliable interpretations and decisions are, since they may include mistakes.

11.2.4 Adjustment of Frequency-Analysis Parameters

The optimal set of frequency-analysis parameters depends on the input. It is generally difficult, in a sound understanding system, to determine a set of parameters appropriate to all possible inputs. It is therefore desirable to adjust these parameters based on the global context which, in turn, is estimated from the previous events provided by the frequency analysis. In the case of beat tracking, appropriate sets of parameters depend on characteristics of the input song, such as its tempo and the number of instruments used in the song.

11.3 OUR APPROACH

Our beat tracking system addresses the general issues discussed in the last section. The following are our main solutions to them.

11.3.1 Multiple Hypotheses Maintained by Multiple Agents

Our way of managing the first issue (ambiguity of interpretation) is to maintain multiple hypotheses, each of which corresponds to a provisional or hypothetical interpretation of the input (Rosenthal, Goto, & Muraoka, 1994; Rosenthal, 1992; Allen & Dannenberg, 1990). A real-time system using only a single hypothesis is subject to garden-path errors. A multiple hypotheses system can pursue several paths simultaneously and decide at later time which one was correct.

BTS is based on multiple-agent architecture in which multiple hypotheses are maintained by programmatic agents which use different strategies for beat-tracking (Figure 11.1 shows the processing model of BTS). Because the input signals are examined according to the various viewpoints with which these agents interpret the input, various hypotheses can emerge. For example, agents

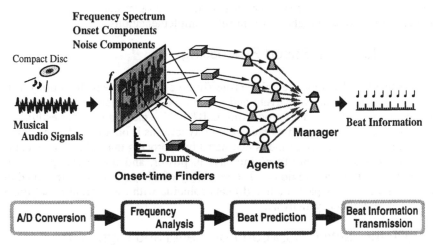

Frequency Spectrum
Onset Components
Noise Components

Compact Disc

Musical
Audio Signals

Drums

Onset-time Finders

Manager

Beat Information

Agents

A/D Conversion ➤ Frequency Analysis ➤ Beat Prediction ➤ Beat Information Transmission

FIG. 11.1. Processing model.

that pay attention to different frequency ranges may predict different beat positions.

The multiple-agent architecture enables BTS to survive difficult beat-tracking situations. Even if some agents lose track of beats, BTS will correctly track beats as long as other agents keep the correct hypothesis. Each agent interprets notes' onset times obtained by frequency analysis, makes a hypothesis, and evaluates its own reliability. The output of the system is then determined on the basis of the most reliable agent.

11.3.2 Musical Knowledge for Understanding Context

To handle the second issue (context-dependent decision), BTS leverages musical knowledge represented as pre-stored drum patterns. In our current implementation, BTS deals with popular music in which drums maintain the beat. Drum patterns are therefore a suitable source of musical knowledge. A typical example is a pattern where a bass drum and a snare drum sound on the strong and weak beats, respectively; this pattern is an item of domain knowledge on how drum-sounds are frequently used in a large class of popular music. Each agent matches such pre-stored patterns with the currently detected drum pattern; the result provides a more global view of the tracking process. These results enable BTS to determine whether a beat is strong or weak and which interbeat interval corresponds to a quarter note.

Although pre-stored drum patterns are effective enough to track beats at the half-note level in the case of popular music that includes drums, we feel that they are inadequate as a representation of general musical knowledge. Higher level knowledge is therefore necessary to deal with other musical genres and to understand music at a higher level in future implementations.

11.3.3 Reliability-Based Processing

Our way of addressing the third issue (imprecision in event detection) is to estimate reliability of every event and hypothesis. The higher the reliability, the greater its importance in all processing in BTS. The method used for estimating the reliability depends on how the event or hypothesis is obtained. For example, the reliability of an onset time is estimated by a process that takes into account such factors as the rapidity of increase in power, and the power present in nearby time-frequency regions. The reliability of a hypothesis is determined on the basis of how its past-predicted beats coincide with the current onset times obtained by frequency analysis.

11.3.4 Interaction Between Low Level and High Level Processing

To manage the fourth issue (adjustment of frequency-analysis parameters), BTS supports interaction between onset-time finders in the low-level frequency analysis and the agents that interpret the results of those finders at a higher level. IPUS (Integrated Processing and Understanding of Signals) (Nawab & Lesser, 1992) also addresses the same issue by structuring the bi-directional interaction between front-end signal processing and signal understanding processes. This interaction enables the system to dynamically adjust parameters so as to fit the current input signals. We implement a simpler scheme, that is, BTS does not have the sophisticated discrepancy-diagnosis mechanism implemented in IPUS.

BTS employs multiple onset-time finders that have different analytical points of view and are tuned to provide different results. For example, some finders may detect onset times in different frequency ranges, and some may detect with different levels of sensitivity (Figure 11.1). Each of these finders communicates with two agents called an *agent-pair*. Each agent-pair receives onset times from the corresponding finder, and can, in turn, re-adjust the parameters of the finder based on the reliability estimate of the hypotheses maintained by its agents. If the reliability of a hypothesis remains low for a long time, the agent tunes the corresponding onset-time finder so that parameters of the finder are close to these of the most reliable finder-agent pair. In other words, there is feedback between the (high-level) beat-prediction agents and the (low-level) onset-time finders.

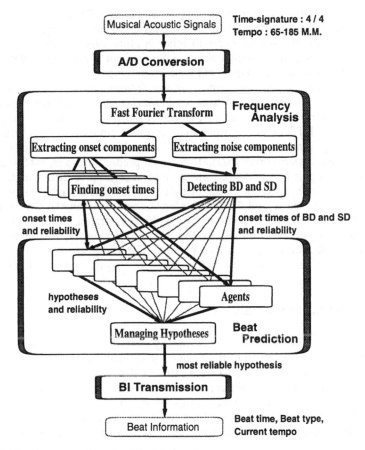

FIG. 11.2. Overview of our beat tracking system.

11.4 SYSTEM DESCRIPTION

Figure 11.2 shows the overview of our beat tracking system. BTS assumes that the time-signature of an input song is 4/4, and its tempo is constrained to be between 65 M.M.[3] and 185 M.M. and almost constant; these assumptions fit a large class of popular music. The emphasis in our system is on finding the temporal positions of quarter notes in audio signals rather than on tracking tempo changes; in the repertoire with which we are concerned, tempo variation is not a major factor. In our current implementation, BTS can only deal with music in which drums maintain the beat. BTS transmits *beat information* (BI) that is the result of tracking beats to other applications in time to the input music. BI consists of the temporal position of a beat (*beat time*), whether the beat is strong or weak (*beat type*), and the current tempo.

The two main stages of processing are *frequency analysis*, in which a variety of cues are detected, and *beat prediction*, in which multiple hypotheses of beat positions are examined in parallel (Figure 11.2). In the frequency analysis stage, BTS detects events such as onset times in several different frequency ranges, and onset times of two different kinds of drum-sounds: a bass drum (*BD*) and a snare drum (*SD*). In the beat prediction stage, BTS manages multiple agents that interpret these onset times according to different strategies and make parallel hypotheses. Each agent first calculates the interbeat interval; it then predicts the next beat time, and infers its beat type, and finally evaluates the reliability of its own hypothesis. BI is then generated on the basis of the most reliable hypothesis. Finally, in the *BI transmission* stage, BTS transmits BI to other application programs via a computer network.

The following describe the main stages of frequency analysis and beat prediction.

11.4.1 Frequency Analysis

Multiple onset-time finders detect multiple tracking cues. First, onset components are extracted from the frequency spectrum calculated by the Fast Fourier Transform. Second, onset-time finders detect onset times in different frequency ranges and with different sensitivity levels. In addition, another drum-sound finder detects onset times of drum-sounds by acquiring the characteristic frequency of the bass drum (BD) and extracting noise components for the snare drum (SD). These results are sent to agents in the beat prediction stage.

Fast Fourier Transform (FFT). The frequency spectrum (the power spectrum) is calculated with the FFT using the Hanning window. Each time the FFT is applied to the digitized audio signal, the window is shifted to the next frame. In our current implementation, the input signal is digitized at 16bit/22.05kHz, the size of the FFT window is 1024 samples (46.44msec), and the window is shifted by 256 samples (11.61msec). The frequency resolution is consequently 21.53Hz and the time resolution is 11.61msec.

Extracting Onset Components. Frequency components whose power has been rapidly increasing are extracted as onset components. The onset components and their degree of onset (rapidity of increase in power) are obtained from the frequency spectrum. The frequency component $p(t,f)$ that fulfills the conditions in (1) is regarded as the onset component (Figure 11.3).

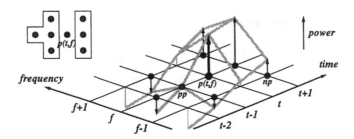

FIG. 11.3. Extracting an onset component.

$$\begin{cases} p(t,f) > pp \\ np > pp \end{cases}$$

EQ. 11.1

Where $p(t,f)$ is the power of the spectrum of frequency f at time t, pp and np are given by:

$$pp = \max(p(t-1,f), p(t-1, f \pm 1), p(t-2,f))$$

EQ. 11.2

$$np = \min(p(t+1,f), p(t+1, f \pm 1))$$

EQ. 11.3

If $p(t,f)$ is an onset component, its degree of onset $d(t,f)$ is given by:

$$d(t, f) = \max(p(t, f), p(t+1)) - pp$$

EQ. 11.4

Finding Onset Times. Multiple onset-time finders[4] use different sets of frequency-analysis parameters. Each finder corresponds to an agent-pair and sends its onset information to the two agents that form the agent-pair (Figures 11.1 & 11.6).

Each onset time and its reliability are obtained as follows: The onset time is given by the peak time found by peak-picking in $D(t)$ along the time axis, where $D(t)$, the sum of the degree of onset, is defined as:

$$D(t) = \sum_{f} d(t,f)$$

EQ. 11.5

$D(t)$ is linearly smoothed with a convolution kernel before its peak time and peak value are calculated. The reliability of the onset time is obtained as the ratio of its peak value to the recent local-maximal peak value.

Each finder has two parameters: The first parameter, *sensitivity*, is the size of the convolution kernel used for smoothing. The smaller the size of the

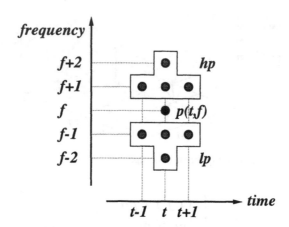

FIG. 11.4. Extracting a noise component.

convolution kernel, the higher its sensitivity. The second parameter, *frequency range*, is the range of frequency for the summation of $D(t)$ (in Equation 11.5). Limiting the range makes it possible to find onset times in several different frequency ranges. The settings of these parameters vary from finder to finder.

Extracting Noise Components. BTS extracts noise components as a preliminary step to detecting SD. Because non-noise sounds typically have harmonic structures and peak components along the frequency axis, frequency components whose power is roughly uniform locally are extracted and considered to be potential SD sounds.

The frequency component $p(t,f)$ that fulfills the conditions in Equation 11.6 is regarded as a potential SD component $n(t,f)$ (Figure 11.4).

$$\begin{cases} hp > p(t,f)/2 \\ lp > p(t,f)/2 \end{cases}$$

EQ. 11.6

$$hp = (p(t \pm 1, f + 1) + p(t, f + 1) + p(t, f + 2))/4$$

EQ. 11.7

$$lp = (p(t \pm 1, f - 1) + p(t, f - 1) + p(t, f - 2))/4$$

EQ. 11.8

Detecting BD and SD. The bass drum (BD) is detected from the onset components and the snare drum (SD) is detected from the noise components. These results are sent to all agents in the beat prediction stage.

(*Detecting onset times of BD*). Because the sound of BD is not known in advance, BTS learns the characteristic frequency of BD that depends on the current song by examining the extracted onset components. For times at which

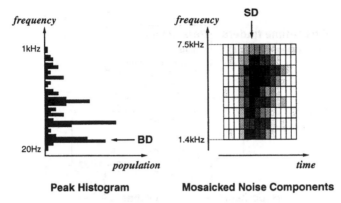

FIG. 11.5. Detecting BD and SD.

onset components are found, BTS finds peaks along the frequency axis and histograms them (Figure 11.5). The histogram is weighted by the degree of onset $d(t,f)$. The characteristic frequency of BD is given by the lowest-frequency peak of the histogram.

BTS judges that BD has sounded at times when (1) an onset is detected and (2) the onset's peak frequency coincides with the characteristic frequency of BD. The reliability of the onset times of BD is obtained as the ratio of $d(t,f)$ currently under consideration to the recent local-maximal peak value.

(*Detecting onset times of SD*). Since the sound of SD typically has noise components widely distributed along the frequency axis, BTS needs to detect such components. First, the noise components $n(t,f)$ are mosaicked (Figure 11.5): the frequency axis of the noise components is divided into subbands[5], and the mean of the noise components in each subband is calculated.

Second, BTS calculates how widely noise components are distributed along the frequency axis $(c(t))$ in the mosaicked noise components: $c(t)$ is calculated as the product of all mosaicked components within middle-frequency range[6] after they are clipped with a dynamic threshold.

Finally, the onset time of SD and its reliability are obtained by peak-picking of $c(t)$ in the same way as in the onset-time finder.

11.4.2 Beat Prediction

To track beats in real time, it is necessary to predict future beat times from the onset times obtained previously. By the time the system finishes processing a sound in an acoustic signal, its onset time has already passed.

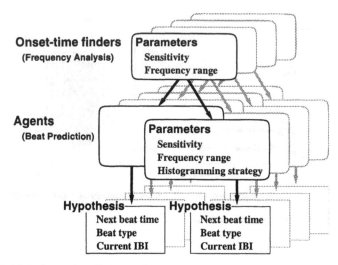

FIG. 11.6. Onset-time finders and agents.

Multiple agents interpret the results of the frequency analysis stage according to different strategies, and maintain their own hypotheses, each of which consists of a predicted next-beat time, its beat type, and the current interbeat interval (*IBI*) (Figure 11.6). These hypotheses are gathered by the manager (Figure 11.1), and the most reliable one is selected as the output.

All agents[7] are grouped into pairs. Two agents in the same pair use the same IBI, and cooperatively predict the next beat times, the difference of which is half the IBI. This enables one agent to track the correct beats even if the other agent tracks the middle of the two successive correct beats (which covers for one of the typical tracking errors). Each agent-pair is different in that it receives onset information from a different onset-time finder (Figure 11.6).

Each agent has three parameters that determine its strategy for making the hypothesis. Both agents in an agent-pair have the same setting of these parameters. The settings of these parameters vary from pair to pair. The first two parameters are *sensitivity* and *frequency range*. These two control the corresponding parameters of the onset-time finder, and adjust the quality of the onset information that the agent receives. An agent-pair with high sensitivity tends to have a short IBI and be relatively unstable, and one with low sensitivity tends to have a long IBI and be stable. The third parameter, *histogramming strategy*, takes a value of either *successive* or *alternate*. When the value is *successive*, successive onsets are used in forming the inter-onset interval (*IOI*)[8] histogram; likewise, when the value is *alternate*, alternate values are used.

The following describe the formation and management of hypotheses. First, each agent calculates the IBI and predicts the next beat time, and then evaluates its own reliability (*Predicting next beat*). Second, the agent infers its

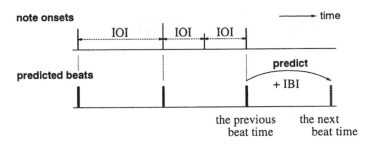

FIG. 11.7. Beat prediction.

beat type sand modifies its reliability (*Inferring beat type*). Third, an agent whose reliability remains low for a long time changes its own parameters (*Adjusting parameters*). Finally, the most reliable hypothesis is selected from the hypotheses of all agents (*Managing hypotheses*).

Predicting Next Beat. Each agent predicts the next beat time by adding the current IBI to the previous beat time (Figure 11.7). The IBI is given by the interval with the maximum value in the inter-onset interval (IOI) histogram that is weighted by the reliability of onset times (Figure 11.8). In other words, the IBI is calculated as the most frequent interval between onsets that have high reliability. Before the agent adds the IBI to the previous beat time, the previous beat time is adjusted to its nearest onset time if they almost coincide.

Each agent evaluates the reliability of its own hypothesis. This is determined on the basis of how the past-predicted beats coincide with onset times. The reliability is increased if an onset time coincides with the beat time predicted previously. If an onset time coincides with a time that corresponds to the position of an eighth note or a sixteenth note, the reliability is also slightly increased. Otherwise, the reliability is decreased.

Inferring Beat Type. Our system, like human listeners, utilizes BD and SD as principle clues to the location of strong and weak beats. Note that BTS cannot simply use the detected BD and SD to track the beats, because the drum detection process is too noisy. The detected BD and SD are used only to label each predicted beat time with the beat type (strong or weak).

Each agent determines the beat type by matching the pre-stored drum patterns of BD and SD with the currently detected drum pattern. The beginning of the best-matched pattern indicates the position of the strong beat.

Figure 11.9 shows two examples of the pre-stored patterns. These patterns represent how BD and SD are typically played in rock and pop music. The beginning of a pattern should be the strong beat, and the length of the pattern is restricted to a half note or a measure. In the case of a half note, patterns repeated twice are considered to form a measure.

The beat type and its reliability are obtained as follows: (1) The onset times of drums are formed into the currently detected pattern, with one sixteenth note resolution that is obtained by interpolating between successive beat times

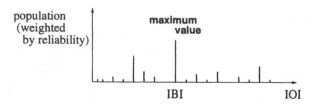

FIG. 11.8. IOI histogram.

(Figure 11.10). (2) The *matching score* of each pre-stored pattern is calculated by matching the pattern with the currently detected pattern: The score is weighted by the product of the weight in the pre-stored pattern and the reliability of the detected onset. (3) The beat type is inferred from the position of the strong beat obtained by the best-matched pattern (Figure 11.11): The reliability of the beat type is obtained from the highest matching score.

The reliability of each hypothesis is modified on the basis of the reliability of its beat type. If the reliability of the beat type is high, the IBI in the hypothesis can be considered to correspond to a quarter note. In that case, the reliability of the hypothesis is increased so that a hypothesis with an IBI corresponding to a quarter note is likely to be selected.

Adjusting Parameters. When the reliability of a hypothesis remains low for a long time, the agent suspects that its parameter set is not suitable for the current input. In that case, the agent adjusts its parameters cooperatively, that is, considering the states of other agents.

The adjustment is made as follows: (1) If the reliability remains low for a long time, the agent requests permission from the manager to change the parameters. (2) If the reliability of the other agent in the same agent-pair is not low, the manager refuses to let the agent change its parameters. (3) The manager permits the agent to change if it has the lowest sum of the reliability in its agent-pair. The manager then inhibits other agents from changing for a certain period. (4) The agent, having received permission, selects a new set of the three parameters that determine its strategy. If we think of the three parameters forming a three-dimensional parameter space, the agent selects a point that is not occupied by other agents and is close to the point corresponding to the parameters of the most reliable agent. The parameter change then affects the corresponding onset-time finder.

Managing Hypotheses. The manager classifies all agent-generated hypotheses into groups, according to beat time and IBI. Each group has an overall reliability, given by the sum of the reliability of the group's hypotheses. The most reliable hypothesis in the most reliable group is selected as the output and sent to the BI transmission stage.

The beat type in the output is updated only using the beat type that has the high reliability. When the reliability of a beat type is low, its beat type is determined from the previous reliable beat type based on the alternation of

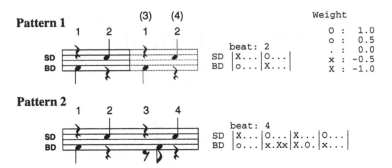

FIG. 11.9. Examples of pre-stored drum patterns.

strong and weak beats. This enables BTS to disregard an incorrect beat type that is caused by some local irregularity of rhythm.

11.5 IMPLEMENTATION

To perform a computationally-intensive task such as processing and understanding complex audio signals in real time, parallel processing provides a practical and realizable solution. BTS has been implemented on a distributed-memory parallel computer, the Fujitsu AP1000 (Ishihata, Horie, Inano, Shimizu, & Kato, 1991) that consists of 64 cells[9].

We apply four kinds of parallelizing techniques (Goto & Muraoka, 1995) to simultaneously execute the heterogeneous processes described in the last section.

11.6 EXPERIMENTS AND RESULTS

We tested BTS on 42 popular songs in the rock and pop music genres. The input was a monaural audio signal sampled from a commercial compact disc, in which drums maintained the beats. Their tempi ranged from 78 M.M. to 184 M.M. and were almost constant.

In our experiment with 8 pre-stored drum patterns, BTS correctly tracked beats in 40 out of 42 songs in real time. At the beginning of each song, beat type was not correctly determined even if the beat time was obtained. This is because BTS had not yet acquired the characteristic frequency of BD. After the BD and SD had sounded stably for a few measures, the beat type was obtained correctly.

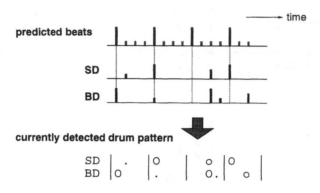

FIG. 11.10. A drum pattern detected from an input.

We discuss the reason why BTS made mistakes in two of the songs. In both of them, BTS tracked only the weak beat, in other words, the output IBI was double the correct IBI. In one song, the number of agents that held the incorrect IBI was greater than that for the correct one. Because the characteristic frequency of BD was not acquired correctly, drum patterns were not correctly matched and the hypothesis with the correct IBI was not selected. In the other song, there was no agent that held the correct IBI. The peak corresponding to the correct IBI in the IOI histogram was not the maximum peak, since onset times on strong beats were often not detected, and an agent was therefore liable to histogram the interval between SDs.

These results show that BTS can deal with realistic musical signals. Moreover, we have developed an application with BTS that displays a computer graphics dancer whose motion changes with musical beats in real time (Goto & Muraoka, 1994). This application has shown that our system is also useful in various multimedia applications in which human-like hearing ability is desirable.

11.7 DISCUSSION

Various beat-tracking related systems have been undertaken in recent years. Most beat tracking systems have great difficulty to work in realistic acoustic environments, however. Most of these systems (Dannenberg & Mont-Reynaud, 1987; Desain & Honing, 1989; Allen & Dannenberg, 1990; Rosenthal, 1992) have dealt with MIDI as their input. Since it is almost impossible to obtain complete MIDI-like representations of audio signals that include various sounds,

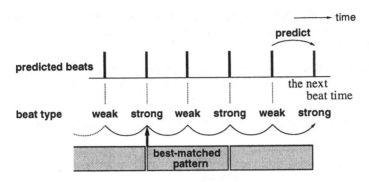

FIG. 11.11. Inferring beat type.

MIDI-based systems cannot immediately be applied to complex audio signals. Although some systems (Schloss, 1985; Katayose, Kato, Imai, & Inokuchi, 1989) dealt with audio signals, they were not able to process music played on ensembles of a variety of instruments, especially drums, and did not work in real time.

Our strategy of first building a system that works in realistic complex environments, and then upgrading the ability of the system, is related to the scaling up problem (Kitano, 1993) in the domain of artificial intelligence. As Hiroaki Kitano stated, "experiences in expert systems, machine translation systems, and other knowledge-based systems indicate that scaling up is extremely difficult for many of the prototypes" (Kitano, 1993). In other words, it is hard to scale up the system whose preliminary implementation works not in real but only laboratory environments. We can expect that computational auditory scene analysis would have similar scaling up problems. We believe that our strategy addresses this issue.

The concepts of our solutions could be applied to other perceptual problems, such as more general auditory scene analysis and vision understanding. The concept of multiple hypotheses maintained by multiple agents is one possible solution in dealing with ambiguous situations in real time. Context-dependent decision making using domain knowledge is necessary for all higher-level processing in perceptual problems. We think reliability-based processing is essential, not only to various processing dealing with realistic complex signals, but to hypothetical processing of interpretations or symbols. As Nawab and Lesser (1992) describe, the mechanism of bi-directional interaction between low-level signal processing and higher-level interpretation has the advantage of adjusting parameter values of the system dynamically to fit a current situation. We plan to apply our solutions to other real-world perceptual domains.

Our beat-tracking model is based on multiple-agent architecture (Figure 11.1) where multiple agents with different strategies interact through competition and cooperation to examine multiple hypotheses in parallel. Although several concepts of the term *agents* have been proposed (Minsky,

1986; Maes, 1990; Nakatani, Okuno, & Kawabata, 1994), in our terminology, the term *agent* means a software component that satisfies the following requirements:

(1) the agent has ability to evaluate its own behavior (in our case, hypotheses of beats) on the basis of a situation of real-world input (in our case, the input song).

(2) the agent cooperates with other agents to perform a given task (in our case, beat tracking).

(3) the agent adapts to the real-world input by dynamically adjusting its own behavior (in our case, parameters).

11.8 CONCLUSION

We have described the main acoustic beat-tracking issues and solutions implemented on our real-time beat tracking system (*BTS*). BTS tracks beats in audio signals that contain sounds of various instruments that include drums, and reports beat information corresponding to quarter notes in time to input music. The experimental results show that BTS can track beats in complex audio signals sampled from compact discs of popular music.

BTS manages multiple agents that track beats according to different strategies in order to examine multiple hypotheses in parallel. This enables BTS to follow beats without losing track of them, even if some hypotheses become incorrect. The use of drum patterns pre-stored as musical knowledge makes it possible to determine whether a beat is strong or weak and which note-value a beat corresponds to.

We plan to upgrade our beat-tracking model to understand music at a higher level and to deal with other musical genres. Future work will include a study on appropriate musical knowledge for dealing with musical audio signals, improvement of interaction among agents and between low-level and high-level processing, and application to other multimedia systems.

11.9 ACKNOWLEDGMENTS

We thank David Rosenthal and anonymous reviewers for their helpful comments on earlier drafts of this chapter. We also thank Fujitsu Laboratories Ltd. for use of the AP1000.

REFERENCES

Allen, P. E. & Dannenberg, R. B. (1990). Tracking musical beats in real time. In *Proc. of the 1990 Intl. Computer Music Conf.*, pp. 140-143.

Dannenberg, R. B. & Mont-Reynaud, B. (1987). Following an improvisation in real time. *In Proc. of the 1987 Intl. Computer Music Conf.*, pp. 241-248.

Desain, P. & Honing, H. (1989).   The quantization of musical time: a connectionist approach. *Computer Music Journal*, 13 (3), 56-66.

Goto, M. & Muraoka, Y. (1994). A beat tracking system for acoustic signals of music. In *Proc. of the Second ACM Intl. Conf. On Multimedia*, pp. 365-372.

Goto, M. & Muraoka, Y. (1995). Parallel implementation of a real-time beat tracking system - real-time musical information processing on AP1000 - (in Japanese). In *Proc. of the 1995 Joint Symposium on Parallel Processing*.

Ishihata, H., Horie, T., Inano, S., Shimizu, T., & Kato, S. (1991).   An architecture of highly parallel computer AP1000. In *IEEE Pacific Rim Conf. on Communications, Computers, Signal Processing*, pp. 13-16.

Katayose, H., Kato, H., Imai, M., & Inokuchi, S. (1989).   An approach to an artificial music expert. In *Proc. of the 1989 Intl. Computer Music Conf.*, pp. 139-146.

Kitano, H. (1993). Challenges of massive parallelism. In *Proc. of IJCAI-93*, pp. 813-834.

Maes, P. (Ed.). (1990). *Designing Autonomous Agents: Theory and Practice from Biology to Engineering and Back*. Cambridge, MA: The MIT Press.

Minsky, M. (1986). *The Society of Mind*. New York: Simon & Schuster, Inc.

Nakatani, T., Okuno, H. G., & Kawabata, T. (1994).   Auditory stream segregation in auditory scene analysis. In *Proc. of AAAI-94*, pp. 100-107.

Nawab, S. H. & Lesser, V. (1992). Integrated processing and understanding of signals. In Oppenheim, A. V., & Nawab, S. H. (Eds.), *Symbolic and*

Knowledge-Based Signal Processing, pp. 251-285. Englewood Cliffs, NJ: Prentice Hall.

Rosenthal, D. (1992).   *Machine Rhythm: Computer Emulation of Human Rhythm Perception*. Ph.D. thesis, Massachusetts Institute of Technology.

Rosenthal, D., Goto, M., & Muraoka, Y. (1994).   Rhythm tracking using multiple hypotheses. In *Proc. of the 1994 Intl. Computer Music Conf.*, pp. 85-87.

Schloss, W. A. (1985). *On The Automatic Transcription of Percussive Music - From Acoustic Signal to High-Level Analysis*. Ph.D. thesis, CCRMA, Stanford University.

NOTES

[1] In this chapter, a *strong beat* is either the first or third quarter note in a measure; a *weak beat* is the second or fourth quarter note.

[2] The *interbeat interval* is the temporal difference between two successive beats.

[3] 65 quarter notes per minute

[4] In the current BTS, the number of onset-time finders is 15.

[5] In the current BTS, the number of subbands is 16.

[6] The current BTS multiplies mosaicked components that are approximately ranged from 1.4kHz to 7.5kHz.

[7] In the current BTS, the number of agents is 30.

[8] The inter-onset interval is the temporal difference between two successive onsets.

[9] A *cell* means a processing element, which has a 25Mhz SPARC with an FPU and 16Mbytes DRAM.

12

Knowledge-Based Analysis of Speech Mixed With Sporadic Environmental Sounds

S. Hamid Nawab, Carol Y. Espy-Wilson, Ramamurthy Mani and Nabil N. Bitar
Boston University

We present the major results of our research in the area of speech recognition in the presence of sporadic environmental sounds. We refer to this general problem area as environmental speech recognition (ESR). Our ESR-related research, which took the form of several case studies, has led to the development of a unified terminology that can be used to describe acoustic scenarios which contain both speech and environmental sounds. We have also used the case studies to examine the viability of a novel knowledge-based approach to ESR problems. The framework for Integrated Processing and Understanding of Signals (IPUS) has been identified as a suitable system architecture for implementing the knowledge-based approach.

12.1 INTRODUCTION

We considered various issues associated with the problem of environmental speech recognition (ESR). Our investigation of ESR issues took the form of several case studies involving spoken telephone numbers and environmental sounds including telephone rings, knocks, clinks, and a car-passing-by sound. Recognition of the spoken digits was required to be expressed in terms of their constituent manner-of-articulation classes - vowels, fricatives, sonorant consonants, and stops. In carrying out these case studies, we formulated the following major conclusions:

(1) Precise formulation of ESR problems requires the development of a **unified terminology** for the acoustic description of scenarios containing both speech and environmental sounds.

(2) A promising **approach to ESR problems** is to formulate knowledge-based solutions which involve the identification of regions in the time-frequency plane that have high speech-to-noise ratios.

(3) ESR solutions can be incrementally improved by using **increasingly source-specific knowledge** about speech and environmental sounds to guide the extraction of information from acoustic data.

(4) The **IPUS framework** (Nawab & Lesser, 1992; Lesser et al., 1993) offers a promising system architecture for solving ESR problems.

We present here some of the major points arising out of our case studies in support of each of the above conclusions. A detailed description of one of our case studies is also presented.

12.2 UNIFIED TERMINOLOGY

At the onset of our ESR-related research, a unified terminology was not available for describing *both* speech and environmental sounds. We utilized existing terminology for classifying speech sounds to devise a generic (speech independent) framework for describing sounds. Furthermore, we used this generic framework to build a new terminology for describing environmental sounds. As part of this terminology, we defined the concept of *acoustons* to describe nonsilent portions of environmental sounds. This concept is analogous to describing nonsilent portions of speech in terms of manner-of-articulation classes.

The benefits that resulted from defining the unified terminology may be summarized as follows:

(1) We defined 24 classes of acoustons for describing segments of environmental sounds. The classification of these acoustons is organized in accordance with a hierarchy of acoustic properties such as rising energy, falling energy, narrowband, wideband, and impulsive.

(2) We used the acouston terminology to formulate specific acoustic models for the environmental sounds used in our case studies. For example, the telephone ring was described as a sequence of 3 different classes of

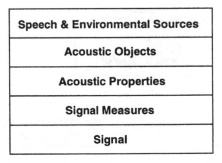

Speech & Environmental Sources
Acoustic Objects
Acoustic Properties
Signal Measures
Signal

FIG. 12.1. Hierarchy of terms used to describe sounds. At the highest abstraction level, multiple-source scenarios are described in terms of their constituent speech and environmental sources. The nonsilent regions of each source are described as sequences of distinct acoustic objects. An acoustic object is described by a group of acoustic properties. Various signal measures are used to gather evidence for the presence of acoustic properties in signal data corresponding to multiple-source scenarios.

acoustons. The description of the entire set of sounds used in our case studies required a total of 8 different classes of acoustons.

(3) We used the unified terminology in our case studies to specify the interactions between speech and environmental sounds. For instance, the terminology allowed us to conveniently express the acoustic phenomena that occur when an environmental sound such as a telephone ring occurs simultaneously with speech manner classes.

12.2.1 Generic

In this section, we introduce a generic terminology that can be used to describe scenarios that contain sounds (speech and environmental) from multiple sources. The hierarchy of this generic terminology is illustrated in Figure 12.1. At the highest abstraction level, a scenario is described in terms of its constituent speech and environmental sound *sources*, and the time regions over which these sources are active. Each sound source is then segmented into nonsilent subregions. A nonsilent subregion is described as a sequence of disjoint *acoustic objects* which have distinct *acoustic properties*. Evidence for the presence of such acoustic properties is extracted from acoustic data corresponding to the multiple-source scenario by using various *signal measures*.

In Figure 12.2, we show an example scenario that is described using our generic terminology. This scenario contains the sound of a ringing telephone and a speech utterance "28." The ringing telephone sound is broken up into nonsilent temporal subregions that correspond to individual rings. An individual ring consists of a sequence of three disjoint acoustic objects. They are classified as "ring attack," "ring regular," and "ring decay." Each of these acoustic objects is described in terms of acoustic properties such as narrowband, rising energy or

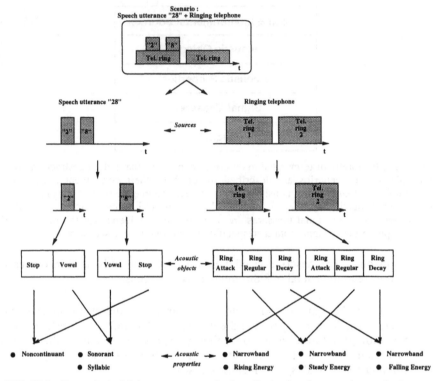

FIG. 12.2. Example multiple-source scenario described using the generic terminology.

falling energy. On the other hand, the speech utterance "28" has nonsilent temporal subregions corresponding to the digits "2" and "8." These subregions are described in terms of a sequence of acoustic objects derived from speech manner classes (vowels, stops, sonorant consonants, fricatives). We give these acoustic objects the same names as the corresponding speech manner classes. The acoustic objects are described in terms of acoustic properties corresponding to linguistic characteristics such as sonorant, syllabic, and noncontinuant.

The generic terminology described here formed the basis for defining the terminology related to environmental sounds that is presented in section 12.2.3. Furthermore, our generic terminology was useful in describing the multiple source scenarios encountered in our case studies.

12.2.2 Speech

In all our ESR case studies, spoken digit sequences constituted the speech part of any acoustic scenario. We refer to a spoken digit sequence as a "speech source". Each speech source is described in terms of subregions which correspond to the individual digits. A spoken digit is in turn described as a

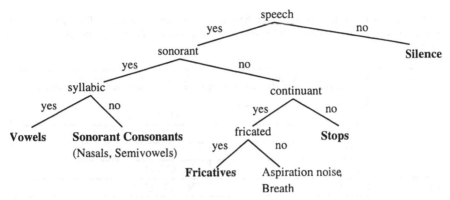

FIG. 12.3. Tree illustrating the relationship between manner classes and linguistic features.

sequence of acoustic objects. We define these acoustic objects to be the acoustic manifestations of the linguistically defined manner-of-articulation classes. Each manner-of-articulation class is obtained by grouping together all phonemes which are similarly produced. For instance, all vowels (e.g., /a/, /i/, /I/, etc.) are produced with a periodic source at the glottis and a vocal tract that is relatively unconstricted, while all fricatives (e.g., /s/, /θ/, /f/, /z/, etc.) are generated with a dominant noise source at a constriction somewhere in the vocal tract. For our research, we utilized four manner-of-articulation classes to describe all speech sounds: vowels, sonorant consonants, stops, and fricatives. Each of these manner classes is characterized by a subset of the following *linguistic features*: sonorant, syllabic, continuant, and fricated. These linguistic features are used to characterize manner classes because they describe the degree of openness or closure of the vocal tract.

The tree shown in Figure 12.3 shows the hierarchy that (1) describes the dependence among linguistic features and (2) depicts different description levels for manner classes. The details of the description become more manner-class-specific as the tree is traversed from the root node to the terminal nodes. For instance, vowels and sonorant consonants are grouped together as "sonorant," but they are separated at the more detailed "syllabic" description level. Furthermore, each of these linguistic features has associated acoustic properties that can, at present, be detected more robustly than other properties such as those related to the place of articulation (labial, alveolar, velar, etc.). Therefore, the acoustic properties related to each of these linguistic features are used to describe manner-class acoustic objects. For instance, the "vowel" acoustic object is described by the "sonorant" acoustic property and the "syllabic" acoustic property. The evidence for each acoustic property is gathered through measurements made on the signal data.

Table 12.1 A list of the 10 digits and their corresponding phonetic spellings and manner-class sequences. In this table, VO = vowel, FR = fricative, ST = stop and SC = sonorant consonant

Digit	Phonetic spelling	Manner-class sequence
oh	o^w	VO
zero	$z\ i^y\ r\ o^w$	FR VO SC VO
one	$w \land n$	SC VO SC
two	$t\ u$	ST VO
three	$\theta\ r\ i^y$	FR SC VO
four	$f \circ r$	FR VO SC
five	$f\ a^y\ v$	FR VO FR
six	$s\ \imath\ k\ s$	FR VO ST FR
seven	$s\ \varepsilon\ v\ \dotlessi\ n$	FR VO FR VO SC
eight	$e^y\ t$	VO ST
nine	$n\ a^y\ n$	SC VO SC

In our ESR research, we have considered the task of recognizing spoken telephone numbers in terms of manner-class acoustic objects. Table 12.1 shows the phonetic spelling and manner-class sequence for each of the 10 digits. The recognition of spoken telephone numbers was performed in terms of manner classes due to the following reasons: (1) each digit can be written as a distinct sequence of the manner classes (see Table 12.1) with the exception of the two digits "9" and "1"[1] and (2) the acoustic properties used to describe manner-class acoustic objects can be robustly detected (Zue, 1985).

12.2.3 Environmental Sounds

Precise formulation of ESR problems requires a systematic description of the acoustic interaction between environmental sources and speech sources. A systematic description of this interaction necessitates a framework for building acoustic specifications of environmental sources. In section 12.2.1, we introduced the concept of *acoustic objects* to describe nonsilent subregions of speech and environmental sources. In the case of environmental sounds, we refer to these acoustic objects as acoustons.

Nonsilent subregions of environmental sounds are described as a sequence of distinct acoustons. These acoustons are distinct in terms of the acoustic properties that are used to describe them. The hierarchy for classification of acoustons is shown in Figure 12.4. A nonsilent subregion of an environmental source may be described in terms of the following four basic types of acoustons: attacking (A), decaying (D), impulsive (I), and regular (R). Such a description

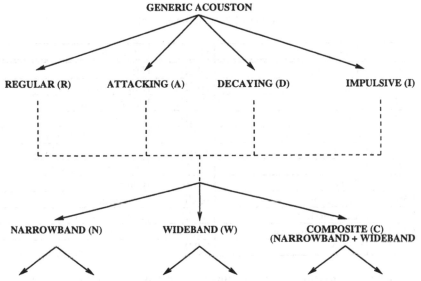

FIG. 12.4. Classification hierarchy of acoustons.

is made on the basis of the temporal variation of energy in the subregion. Each of these acoustons may be further classified into three types on the basis of spectral content of the time-region they describe. These three types are narrowband (N), wideband (W), and composite (C). This classification may be extended to six types on the basis of a more detailed spectral description. For instance, a narrowband acouston may be classified as simple or complex on the basis of the number of spectral components it contains. In Figure 12.4, we also introduce abbreviations for the different acouston types. These abbreviations establish a "language" for describing the nonsilent subregions of environmental sound sources in terms of acoustons.

We found the description framework introduced in this section to be useful in describing a wide variety of environmental sources. In Table 12.2 we present examples of sounds with different types of acoustons. The task of describing the acoustic interaction between speech and environmental sound sources in our case studies was made simpler by the introduction of this terminology.

12.3 ESR APPROACH

We formulated a knowledge-based approach for performing manner-class recognition assuming degraded speech (MCR-ADS). Our approach relies on

Table 12.2 Acouston examples

Type	Example	Type	Example
NSR	Smoke alarm	WHR	Hiss
NSA	Whistle	WHA	Start of hiss
NSD	Siren	WHD	Air escaping tire
NSI	Beep	WHI	Scrape
NCR	Telephone ring	CMR	Hair-dryer kept on
NCA	Burglar alarm	CMA	Hair-dryer turned-on
NCD	Musical note	CMD	Hair-dryer turned-off
NCI	Clink	CMI	Clang
WLR	Car interior	CUR	Car-passing-by
WLA	Airplane taking-off	CUA	Car-passing-by
WLD	Airplane landing	CUD	Car-passing-by
WLI	Knock	CUI	Clang

augmenting a procedure that performs manner-class recognition assuming clean speech (MCR-ACS) with additional techniques for identifying the occurrences of environmental sounds. High-level knowledge regarding environmental sounds is then used to identify regions of the time-frequency plane that are least influenced by these sounds. The procedure for performing MCR-ACS analysis is then restricted to regions with a high speech-to-noise ratio to refine or modify the initial manner-class hypotheses. This approach may be viewed as specifying a problem-solving architecture which permits primitive segregation to be closely coordinated with schema-based segregation, as suggested by Bregman (1990).

We demonstrated the applicability of our MCR-ADS approach through three case studies that involved a spoken telephone number contaminated with different environmental sounds. To demonstrate the effectiveness of our MCR-ADS approach, we present here the major recognition results obtained in each of the case studies. To identify recognition errors in the case of contaminated speech, we assumed that the manner-class hypotheses obtained by applying our MCR-ACS procedure to the uncontaminated speech were error-free. The recognition errors were classified into 3 categories: substitutions, insertions, and deletions[2]. The results below show that for each source of contamination, our knowledge-based MCR-ADS procedure significantly reduced the recognition errors with respect to the application of our MCR-ACS procedure to the contaminated speech. The remaining recognition errors (after the application of our MCR-ADS procedure) can be easily eliminated by the application of digit level constraints on allowable sequences of manner-class hypotheses.

(1) When the contamination was due to a telephone ring, the MCR-ACS results contained 10 recognition errors (3 substitutions, 2 deletions, and 5 insertions). The output of our full-blown MCR-ADS procedure contained only 3 recognition errors (0 substitutions, 3 deletions, and 0 insertions).

(2) When the contamination was due to 3 knocks and a clink, the MCR-ACS results contained 6 recognition errors (0 substitutions, 0 deletions, and 6 insertions). All recognition errors were eliminated by our full-blown MCR-ADS procedure.

(3) When the contamination was due to a car-passing-by sound, the MCR-ACS results contained 16 recognition errors (9 substitutions, 4 deletions, and 3 insertions). The output of our full-blown MCR-ADS procedure contained 8 recognition errors (2 substitutions, 5 deletions and 1 insertion). Although many of the remaining errors had been identified, they could not be corrected because of the low speech-to-noise ratios in the corresponding time regions.

12.4 SOURCE-SPECIFIC KNOWLEDGE

During the course of performing MCR-ADS, we found that the quality of ESR solutions may be improved with the use of increasingly source-specific knowledge about both speech and environmental sources. In particular, the source-specificity of a sound model enables the system to identify contaminated time-frequency regions not detected during the initial processing. On the other hand, the availability of a hierarchy of less source-specific models aids the search process for identifying more source-specific source models.

A source model may be conceptualized as consisting of a set of constraints on a corresponding acoustic object. Such a model may be made more source-specific by augmenting its constituent set of constraints. It should be noted that as source-specificity of a model is increased, the set of sources spanned by the model becomes smaller. To illustrate these ideas, let us consider an example of source models of increasing source-specificity as shown in Figure 12.5. The "narrowband" model simply requires that the detected acoustic energy forms narrowband tracks in the time-frequency plane. At a more source-specific level, Figure 12.5 includes a "ringing" model that requires the presence of at least one narrowband component of a particular type between 1 and 2 kHz. The figure also includes more specific models corresponding to a telephone ring and the ring of a particular telephone (Joe's telephone).

To illustrate the advantage of having access to models of increasing source-specificity, consider a situation where a narrowband signal component has been detected at 1.3 kHz during initial processing by a procedure that performs the MCR-ADS task. If the most source-specific model identified on the basis of this data is the "telephone ring" model of Figure 12.5, we would be able to predict additional contamination to exist anywhere in the 1 to 2 kHz band. However, if the model corresponding to Joe's telephone ring (see Figure 12.5) is available,

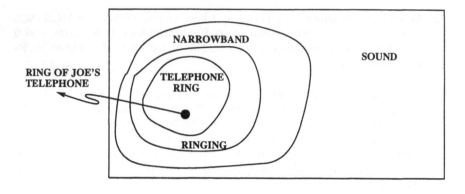

	MODEL	DESCRIPTION
	SOUND	Segment of nonzero acoustic energy
	NARROWBAND	Sound in which energy forms narrowband tracks
Increasing source-specificity	RINGING	Narrowband sound which has at least one component in 1--2 kHz band Narrowband components have onsets at a rate of at least 1 Hz
	TELEPHONE RING	Ringing sound which has at least two narrowband components in 1--2 kHz band Duration is approximately 2.8 secs
	RING OF JOE'S TELEPHONE	Telephone ring which has narrowband components at 1.3 kHz and 1.6 kHz

FIG. 12.5. An example of models with varying degrees of source-specificity.

we would be able to pinpoint the additional narrowband contamination to a narrow region around 1.6 kHz.

12.5 IPUS FRAMEWORK

We have implemented our MCR-ADS approach utilizing a system architecture that integrates the use of various signal processing algorithms and high-level knowledge about speech and environmental sounds within a single problem-solving framework. This architecture, known as IPUS, provides generic mechanisms for (1) detecting discrepancies among the results obtained from an initial processing phase, (2) searching for plausible explanations for those discrepancies and (3) planning and executing a reprocessing phase aimed at producing nondiscrepant results. For our case studies, the initial processing phase involved the use of an MCR-ACS procedure for forming initial manner-class hypotheses as well as an **onset-rate analysis (ORA)** technique. The ORA technique utilizes the **multiband exponential rate operator** (Mani and Nawab, 1995) to detect a set of acoustic properties related to regularity, sharpness and time-frequency distribution of signal onsets. We then detect discrepancies between the acoustic properties associated with the initially hypothesized manner classes and the *entire* set of acoustic properties detected during the initial phase by MCR-ACS analysis and ORA. A search is then conducted for environmental sound models which plausibly explain the presence of nonspeech properties in the acoustic data. This helps to identify the speech-only portions of the time-frequency plane. The procedure for performing MCR-ACS is then restricted to these speech-only regions of the time-frequency plane in order to obtain a revised set of manner-class hypotheses. The overall process may then be repeated until no further improvement is obtained.

We have demonstrated the utility of our IPUS-based MCR-ADS framework through ESR case studies. The major results that were obtained by applying this framework to each case study are as follows:

(1) When the contamination was due to a telephone ring, our IPUS-based procedure detected, explained, and rectified 7 discrepancies.

(2) When the contamination was due to 3 knocks and a clink, our IPUS-based procedure detected, explained, and rectified 10 discrepancies.

(3) When the contamination was due to a car-passing-by sound, our IPUS-based framework detected 18 discrepancies. Out of these 18 discrepancies, all were explained but only 15 were rectified. The framework could not rectify 3 discrepancies because of the very low speech-to-noise ratios in these regions.

12.6 IMPLEMENTATION

The implementation of the IPUS-based MCR-ADS framework requires (1) a procedure for MCR-ACS analysis and (2) algorithms which detect acoustic properties that can be used for discrepancy detection. We have used the B&E procedure developed by Bitar and Espy-Wilson (1995) to perform the MCR-ACS task. The onset-rate analysis technique (ORA) was used for providing discrepancy detection capabilities. We now briefly describe the B&E procedure and the ORA technique.

Table 12.3 Onset-related acoustic properties detected using ORA.

Property Abbrev	Definition
MR,CF,UN	Medium attack rate onsets occurring at a uniform temporal rate across many frequencies (0 - 4 kHz)
MR,HF	Medium attack rate onsets occurring at high frequencies (4 - 8 kHz)
MR,LF,UN	Medium attack rate onsets occurring at a uniform temporal rate at low frequencies (0 - 2 kHz)
MR,LF,NU	Medium attack rate onsets occurring at a non-uniform temporal rate at low frequencies (0 - 2 kHz)
HR,LF	High attack rate onsets occurring at low frequencies (0 - 2 kHz)
HR,MF	High attack rate onsets occurring at mid frequencies (2 - 4 kHz)
HR,HF	High attack rate onsets occurring at high frequencies (4 - 8 kHz)
LR	Low attack rate frequency component

The B&E procedure uses a set of signal processing algorithms that have been designed under the assumption of clean speech. These algorithms perform measurements on acoustic data in order to provide evidence for acoustic properties that can be used to describe manner-class acoustic objects. The acoustic properties are motivated by a theory that describes manner classes in terms of linguistic features (see Figure 12.3). To model the uncertainty in the recognition process, the B&E procedure uses a fuzzy logic framework (De Mori, 1983) to express the degree of evidence for an acoustic property based on the related acoustic measurement values. The theory behind the B&E procedure as well as more details pertinent to its implementation can be found in (Bitar and Espy-Wilson, 1995).

The ORA technique is used to find evidence for a set of onset-related acoustic properties. This technique is based on the multiband exponential rate operator which is used to calculate the "instantaneous exponential rate"[3] of a signal in different spectral bands. The ORA technique suitably thresholds the output of this operator to detect signal onsets with different onset rates (or attack rates). A set of measurements are performed to characterize the regularity and time-frequency distribution of signal onsets which have significant energy. These measurements constitute evidence for the presence of some onset-related acoustic properties. For instance, a set of measurements are carried out to detect onsets with high attack rates which are concentrated in the low frequency region (0 kHz - 2 kHz). A summary of all the onset properties that are detected using our ORA technique is provided in Table 12.3.

The MCR-ADS task was performed by integrating the B&E procedure and ORA technique in an IPUS framework. In the following section, we present results from one of our case-studies in order to illustrate the performance of this IPUS-based MCR-ADS framework.

12.7 RESULTS

Our ESR-related research took the form of three case studies which involved spoken telephone numbers and different environmental sounds. The environmental sounds were chosen such that they included a wide variety of acoustons. The sounds which were used in our case studies were telephone rings, knocks, clinks, and a car-passing-by sound. In this section, we present detailed results from one of the studies, which involved a speech utterance contaminated by a telephone ring.

The test utterance "2858096" was contaminated by a telephone ring with 6 dB speech-to-noise ratio[4]. The telephone ring, which consists of NCA, NCR and NCD acoustons, started 1.3 secs into the speech utterance. In Figures 12.6(a) and 12.6(b), we show the spectrograms of the clean and contaminated speech.

MCR-ACS on contaminated speech: The contaminated speech utterance was analyzed using the B&E procedure for performing manner-class recognition under the assumption of clean speech (MCR-ACS). The MCR-ACS results are illustrated in Figure 12.7. Here, we represent the MCR-ACS outputs as two-dimensional plots in which the horizontal axis corresponds to time and the vertical axis corresponds to the manner classes. The degree of evidence for each manner class is represented at every time instant by one of eleven gray levels. Black represents highest evidence for the presence of a manner class, while white represents absence of a manner class. We show the MCR-ACS results from 1.2 sec to 2.7 sec for the clean speech in Figure 12.7(a), which we also consider as corresponding to the actual manner classes of the clean speech. In Figure 12.7(b), we show the MCR-ACS results for the time region over which the telephone ring overlaps with the speech utterance. A comparison between Figures 12.7(a) and 12.7(b) reveals that 10 manner-class recognition errors were obtained when our MCR-ACS procedure was applied to the contaminated speech. The errors are listed below:

(1) The ring's NCA acouston is interpreted as a stop due to the abrupt onset.

(2) Fricatives are inserted in the silence regions due to the presence of high frequency components in the ring acoustons.

(3) The fricative /z/ in the digit "zero" is replaced by a vowel due to the presence of high energy at low frequencies in ring's NCR acouston.

(4) A sonorant consonant is inserted in "zero" due to energy decrease in the ring's NCR acouston.

(5) There are shifts in vowel and sonorant consonant labels during the sequence "zero-nine" due to changes in the energy of the ring's NCR acouston.

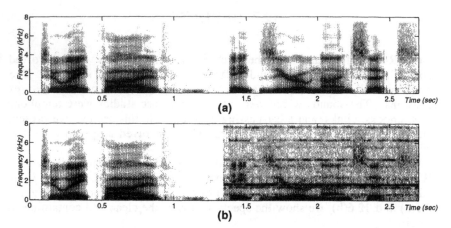

FIG. 12.6. Plot (a) shows the spectrogram of the clean speech utterance "2858096." Plot (b) shows the spectrogram of the utterance "2858096" contaminated with a telephone ring, which starts at approximately 1.3 sec.

(6) Some vowels and sonorant consonants are inserted in fricative regions of "six" due to the low frequency component of the ring's NCR acouston.

MCR-ADS on contaminated speech: The contaminated speech was analyzed using our IPUS-based framework for manner-class recognition assuming degraded speech (MCR-ADS). Two versions of the MCR-ADS task were performed. The first used only a generic source model for the telephone ring, whereas the second version used a more specific source model for the telephone ring. The generic source model described the telephone ring as a sound with narrowband frequency components without specifying the actual locations of the individual components. On the other hand, the specific source model for the telephone ring specified the exact locations of the narrowband frequency components. The results of the first version and second version are shown in Figures 12.8(b) and 12.8(c) respectively. For comparison, we also show the actual manner classification of clean speech in Figure 12.8(a). The MCR-ADS version which used generic source knowledge contained 6 recognition errors, whereas the results of the version which used specific source knowledge contained only 3 recognition errors.

We will now briefly describe how the IPUS-based MCR-ADS framework was used to obtain improved manner-class recognition in the presence of a telephone ring. The B&E procedure for performing the MCR-ACS task and the ORA technique were applied to the contaminated utterance. Discrepancies were detected between the acoustic properties detected by ORA and the expected acoustic properties of the manner class acoustic objects hypothesized by MCR-ACS. For instance, the time region from 1.6-1.7 secs in the MCR-ACS results indicated the presence of a fricative. The ORA output for the same time region did not show any high frequency onsets (denoted by [MR,HF] in Table 12.3) to

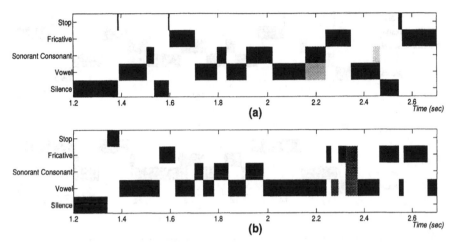

FIG. 12.7. Illustration of manner-class recognition errors obtained when a recognition procedure based on the assumption of clean speech is applied to contaminated speech. Plot (a) shows the actual manner classification for the clean speech of Figure 12.6(a) from 1.2 sec to 2.7 sec. Plot (b) shows the results obtained for the same speech contaminated by a telephone ring. In both plots, the horizontal axis corresponds to time, while the vertical axis corresponds to the manner classes. The degree of evidence for each manner class is represented at every time instant by one of eleven gray levels. Black represents highest evidence for the presence of a manner class, whereas white represents absence of a manner class.

support this result. Therefore a discrepancy was flagged in this time region. The ORA results also showed the presence of low onset-rate (denoted by LR in Table 12.3) components at 1.5 kHz, 1.75 kHz and 4.25 kHz. Since speech formants have to show regular energy onsets at the pitch period, it was concluded that the low onset-rate components could not have arisen from speech sounds. This conclusion was used to hypothesize the presence of a narrowband environmental sound in this scenario. This hypothesis was used to explain most of the discrepancies. The data was reprocessed after excluding the frequency bands corresponding to the low onset-rate components of the environmental sound. Figure 12.8(b) shows MCR-ADS results that were generated after such an exclusion. A comparison of Figure 12.8(b) with Figure 12.8(a) shows that the use of this knowledge was clearly sufficient to correct many of the manner recognition errors that resulted from the presence of the telephone ring. However, there still remain fricative insertions at 1.5 sec and 2.5 sec due to the high frequency component (at 6.5 kHz and undetected by ORA) of the telephone ring. We now illustrate how more specific knowledge about telephone rings can be used to eliminate these errors.

In Figure 12.8(c), we show the results of MCR-ADS with a telephone-ring model which specifies the existence of a high frequency component around 6.5 kHz. The figure clearly shows that the two inserted fricatives were removed and

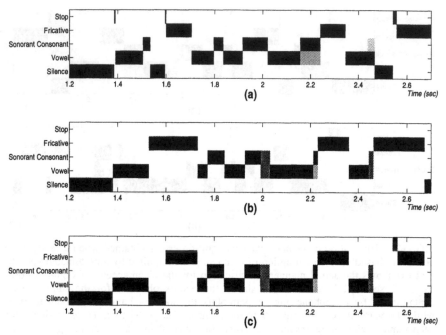

FIG. 12.8. Illustration of improved manner-class recognition for contaminated speech using our knowledge-based approach. Plot (a) shows the actual manner classification for the clean speech. Plot (b) shows the manner class recognition results for contaminated speech when using our knowledge-based approach and limited knowledge about the characteristics of a telephone ring. Plot (c) shows the manner class recognition results for contaminated speech when using detailed knowledge about the characteristics of a telephone ring. Notice that most of the recognition errors of Fig. 12.7(b) have been eliminated in Fig. 12.8(c). The only remaining errors of consequence are located between 1.4 sec and 1.6 sec, and they take the form of two deleted stops and one deleted sonorant consonant.

that one deleted stop (at 2.55 sec) was recovered. Even though the ORA detected only three low onset-rate components at 1.5 kHz, 1.75 kHz and 4.25 kHz, these were sufficient to hypothesize the occurrence of a telephone ring. The telephone-ring model was then used to predict the presence of an additional narrowband component at 6.5 kHz. Consequently, the portion of the B&E procedure which uses a measurement of high-frequency energy to hypothesize the presence of fricatives was made to exclude a band of frequencies around 6.5 kHz. This helped in preventing the fricative insertions that occurred when only generic source knowledge was used for the telephone ring.

12.8 ACKNOWLEDGEMENTS

This work was supported by NSF research grant # IRI-9300194.

REFERENCES

Bitar, N. N., & Espy-Wilson, C. Y. (1995). A signal representation of speech based on phonetic features. In *Proceedings of the 1995 Fifth Annual IEEE Dual-Use Technologies & Applications Conference*, Utica, New York.

Bregman, A. S. (1990). *Auditory Scene Analysis: The Perceptual Organization of Sound.* Cambridge, MA: MIT Press.

De Mori, R. (1983). *Computer Models of Speech Using Fuzzy Algorithm.* New York: Plenum Press.

Lesser, V., Nawab, S. H., Gallastegi, I., & Klassner, F. (1993). IPUS: An architecture for integrated signal processing and signal interpretation in complex environments. In *Proceedings of the 1993 National Conference on Artificial Intelligence (AAAI-93)*, pp. 249-55, Washington, DC.

Mani, R. & Nawab, S. H. (1995). A multiband exponential rate operator for musical transient analysis. In *Proceedings of the 1995 International Conference on Acoustics Speech and Signal Processing*, volume 2, pp. 1233-1236, Detroit, Michigan. IEEE.

Nawab, H. & Lesser, V. (1992). Integrated Processing and Understanding of Signals. In S. H. Nawab, and A. V. Oppenheim (eds.), *Symbolic and Knowledge-Based Signal Processing* (chapter 7). Englewood Cliffs, NJ: Prentice Hall.

Zue, V. (1985). The use of speech knowledge in automatic speech recognition. *Proceedings of the IEEE*, 73(11), 1602-15.

NOTES

---

[1] These two digits can be distinguished if we add the nasal manner class to our inventory. Acoustic properties for this class are currently being studied.

[2] A substitution occurs when one manner-class hypothesis replaces another in a sequence of manner-class hypotheses. A deletion occurs when a manner-class hypothesis is eliminated from a sequence of manner-class hypotheses. An insertion occurs when a new manner-class hypothesis is inserted in a sequence of manner-class hypotheses.

[3] For an exponential signal of the form $x(t) = e^{at}$, we call a the exponential rate. The terms onset-rate, attack-rate and decay-rate are also associated with this parameter. The parameter $\dfrac{1}{a}$ is called the time-constant of the signal.

[4] We define speech-to-noise ratio as:

$$Speech-to-noise\ ratio = 10 \log_{10}\left(\frac{Energy\ in\ speech}{Energy\ in\ noise}\right)$$

The speech and noise energies are computed in the temporal region over which the two signals overlap.

13

Multiagent Based Binaural Sound Stream Segregation

Tomohiro Nakatani & Hiroshi G. Okuno
NTT Basic Research Laboratories

Masataka Goto
Waseda University

Takatoshi Ito
Toyohashi University of Technology

This chapter presents a multiagent based sound stream segregation system for binaural input. Sound stream segregation is a technology used to extract individual sounds from a mixed sound, and it is considered to be primary processing for computationally understanding sounds (Computational Auditory Scene Analysis) in the real world. We already proposed residue-driven architecture as a computational model for a sound stream segregation. In this chapter, we discuss the design and implementation of two subsystems, called *agencies*, the BiHBSS agency and the BiGrouping agency, based on this architecture. The two agencies together segregate sound streams according to their harmonic structure and direction information. First, the BiHBSS agency segregates fragments of streams, and then, the BiGrouping agency groups the fragments to extract streams. Experimental results show that these agencies improve the quality of sound stream segregations.

13.1 INTRODUCTION

In the real acoustical environment where many sounds reach our ears at the same time, we can listen adaptively to a particular sound in the sound mixture by focusing on it. This phenomenon is known as the *cocktail party effect*. Currently, no computer systems have such an adaptive sound understanding mechanism. For constructing a much more friendly human-computer interface in the real world, this adaptivity is very important. Recently, researchers began to study *Computational Auditory Scene Analysis (CASA)* and are trying to construct a general computational framework for understanding the real sound environment.

Because the input of a CASA system is a mixed sound, sound stream segregation is an important primary processing for understanding individual acoustical events. A sound stream is a group of sounds that have some consistent attributes (Bregman, 1990). Several models have been proposed for sound stream segregation. In order to construct a flexible and expandable sound stream segregation system, we proposed a multiagent based approach, called *residue-driven architecture (RDA)* (Nakatani et al., 1995b), in which agents are dynamically generated to extract individual sound streams by focusing on the consistent attributes of the sound streams. Agents interact with each other to extract sound streams exclusively, and the whole system can adaptively segregate individual sound streams. In the RDA, a subsystem for extracting sound streams is called an agency. We have already presented the Harmonics-Based Stream Segregation *(HBSS)* agency that segregates fragments of sound streams based on harmonic structures under noisy conditions (Nakatani et al., 1994a).

Along with harmonic structure, information on the sound location is quite important in stream segregation. In this chapter, we discuss the design and implementation of a *BiHBSS* (binaural harmonics based stream segregation) agency and a *BiGrouping* (binaural grouping) agency. The BiHBSS agency segregates sound fragments from binaural sound input. (We call a sound fragment segregated by BiHBSS agency a *stream fragment* because it is only a part of a sound stream.) The BiGrouping agency extracts sound streams by grouping the output stream fragments of the BiHBSS agency. Both agencies use information on sound direction and harmonic structure as consistent attributes. Experimental results show that these agencies can effectively extract two sound streams from a sound mixture.

The rest of this chapter is organized as follows: Section 13.2 discusses a computational model of sound stream segregation and the residue-driven architecture. Section 13.3 describes the design and implementation of the binaural stream segregation agencies, and Section 13.4 presents experimental results. Section 13.5 is a conclusion.

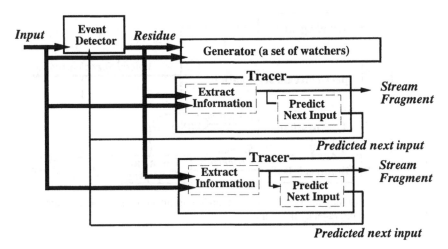

FIG. 13.1.   Residue-Driven Architecture.   The architecture consists of three kinds of agents, an event-detector, a tracer-generator and tracers.   This subsystem is called an *agency*.

13.2 COMPUTATIONAL MODEL OF SOUND STREAM SEGREGATION

13.2.1 Sound Stream Segregation

A sound coming from a particular sound source has some consistent attributes, such as harmonic structure, direction, common AM, common FM, timbre, and loudness.   Because a lot of sounds occur and their mutual influence on each other changes dynamically, there is a need for not only a mechanism for detecting individual sound attributes but also one for operating their dynamic relationships.   Moreover, these sounds usually contain complex information and it is quite difficult to handle all of it at once.   For a system to be able to handle this, we think that the architecture should support a framework for incremental design and implementation (Okuno, 1993, Nakatani et al., 1994b).   Because sound stream segregation systems will be used as building blocks for real-time applications, they should segregate sound streams incrementally.

For constructing a dynamic, adaptive and expandable sound stream segregation system, a multiagent system proposed in artificial intelligence is promising (Minsky, 1986, Maes, 1991).   In this model, intelligence is considered as an overall property that emerges through interactions among many agents.

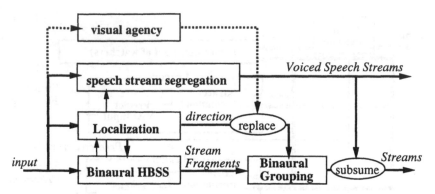

FIG. 13.2. Interagency interaction by subsumption architecture. Extracted streams can be refined by using interagency interaction.

13.2.2 Residue-Driven Architecture

Sound stream segregation systems must 1) find a new stream, 2) trace the stream, 3) detect that the stream has ceased, and 4) resolve any interference between simultaneous streams.

In the RDA for sound stream segregation, a subsystem, or *agency*, extracts sound streams by using sound attributes. Each agency consists of three kinds of agents, an event-detector, a tracer-generator and tracers (Figure 13.1). A tracer-generator detects a new stream, and it generates a tracer. A tracer extracts the stream by continuously tracking its consistent attributes, and predicts its next input and sends it to the event-detector. The event-detector controls the flow of interaction between agents so that each sound components should be allocated to one sound stream exclusively. For this purpose, the event-detector subtracts a set of predicted inputs from the actual input sounds and sends the residual signal, called *residue*, to the tracer-generator and tracers. Through the event-detector, tracers indirectly inhibit other tracers from extracting the same sound components and inhibit the tracer-generator from generating redundant tracers.

In the RDA, the system consists of a set of agencies and it can be expanded by adding new agencies. A simplified scheme showing a set of agencies is depicted in Figure 13.2. One way the agencies interact is through the input/output relation (e.g., the relation between the BiHBSS agency and the BiGrouping agency), and another way is the subsumption architecture (Brooks, 1986) (e.g., the relation between the Localization agency and the BiHBSS agency). For example, the localization agency in Figure 13.2 extracts the spatial information from binaural input, and replaces the direction information of streams for the binaural stream segregation agencies. The visual agency may extract more precise location information and replace it. Thus, the interagency interaction provides a computational mechanism to refine extracted streams with subsumption architecture.

FIG. 13.3. BiHBSS agency and BiGrouping agency. The BiHBSS agency segregates stream fragments, and the BiGrouping agency segregates streams by grouping those fragments.

The binaural stream segregation system with the RDA works as follows (Figure 13.3). The BiHBSS agency receives binaural input and outputs stream fragments. Tracers track continuously changing sound attributes of harmonic structure and direction and extract sound stream fragments. Tracers interact through the event-detector to avoid redundant tracer generation. The BiHBSS agency can only extract a stream fragment whose harmonic structure continues unbroken from start to finish. Some sound sources produce discontinuous harmonic streams in which harmonic stream fragments occur intermittently. For example, a sequence of vowels in speech is usually an intermittent sound stream. To segregate such sound streams, it is necessary to sequentially group stream fragments according to the consistent attributes of the sound sources. The BiGrouping agency receives stream fragments segregated by the BiHBSS agency and sequentially groups them to segregates sound streams according to consistencies in the fundamental frequency and direction. Tracers in the BiGrouping agency exclusively allocate stream fragments to groups in similar to the way the BiHBSS agency does. We designed the above agencies so that they segregate streams, or stream fragments, incrementally since we think this is required for adaptive understanding of sounds in the real-world. The BiHBSS agency extracts stream fragments, while the BiGrouping agency groups the extracted fragments concurrently. The BiHBSS agency deals with local consistency of stream fragments, while the BiGrouping agency deals with more global relationship between them. Because the BiGrouping agency can access more global information than the BiHBSS, the BiGrouping agency can be expected to extract correct attributes of stream fragments more stably than the BiHBSS. As a result, the system adaptively segregate individual stream groups and their component fragments.

13.2.3 Related Works

Many techniques to segregate sound streams has been presented. Some systems use an auditory model for primary processing (Lyon, 1984, Brown, 1992, Slaney et al., 1994). Brown and Cooke presented a computational model with auditory maps for representing features that segregates streams by integrating those sound features. Since the integration process becomes complicated when treating a real, complex sound mixture, blackboard architecture is now being used to simplify this integration process (Cooke, et al., 1993). The auditory map model has some practical limitations. It is an off-line algorithm in the sense that any part of the input is available to the algorithm at any time. However, batch algorithms are poor at providing a real-time application. Additionally, it is not easy to incorporate schema-based stream segregation into such a system, because it does not support a mechanism for extensibility.

Kashino proposed OPTIMA, a computational architecture for integrating the bottom-up and top-down processing of music sound source separation using the hypothesis network (Kashino, et al., 1995). Because this system is a stochastic approach based on Bayes estimation, it is expected to work efficiently if all the conditional probabilities of acoustical events are appropriately given. However, it is very difficult to determine all the probabilities in a real sound environment, and the system may have problems with unknown sounds.

To design a more flexible and expandable system, control mechanisms are needed. IPUS (*Integrated Processing and Understanding Signals*) (Lesser, et al., 1993) integrates signal processing and signal interpretation into the blackboard system. It uses a small set of front-end signal processing algorithms (SPAs) and chooses the correct parameter setting for each SPA and computes the correct interpretation by dynamic SPA reconfiguration. This reconfiguration is viewed as a diagnosis for discrepancy between topdown search for SPA parameter setting and bottomup search for the correct interpretation. IPUS has various interpretation knowledge sources that understand actual sounds, such as hair dryers, footsteps, telephone rings, fire alarms, and waterfalls (Nawab, et al., 1992). However, it may have problems in upgrading, because when the number of SPAs increases it may fail in choosing the correct parameter settings. Additionally, in order to support a reactive response, another system may be needed to compute the required information.

Using interfering signal cancellation, some specific sounds can be eliminated from the input (Costas, 1981, Ramalingam, et al., 1994), and the processing of the other sounds then becomes easy. For example, the Residual Interfering Signal Canceler (RISC) can effectively segregate some sound streams by using specific sound attributes (Ramalingam, et al., 1994). Therefore, it represents a simple and powerful way to resolve stream interference when the attributes of each signal are known. However, the mechanism by which RISC architecture treats the number of sound sources and copes with different attributes of sounds has not been considered in any depth.

With spatial information, sound separation technology using microphone array has been developed (Schmids, 1986, Stadler, et al., 1993). There are some limitations of the array method on the number of sound sources and on the frequency region. Moreover, the technology itself does not provide a sufficient sound stream segregation mechanism when spatial information is not useful because of, for example, more than one sound stream coming from the same direction. Some technology for integrating several sound stream segregation keys with the array method is required.

13.3 DESIGN AND IMPLEMENTATION

13.3.1 System Design

Here, we describe the design and implementation of the BiHBSS agency and the
BiGrouping agency (see Figure 13.3). The BiHBSS agency segregates
harmonic stream fragments with information of sound direction, while the
BiGrouping agency extracts sound streams by grouping those fragments. Each
agency can be modeled by the RDA.

To extract sound direction from binaural sound, many techniques have been
proposed (Bodden, 1993, Lyon, 1984, Schmids, 1986). We use fine structure of
harmonics sound, that is, the interaural time difference (*ITD*) and interaural
intensity difference (*IID*) of overtones, as localization clues. Our system first
extracts harmonic information in both channels, and extracts the direction on the
harmonics. Once the direction information is extracted, the system utilizes it as
a segregation clue for the successive harmonic sound stream. While the HBSS
agency provides the monaural harmonics stream segregation, the BiHBSS
agency provides an advanced stream segregation using direction information.

In the following sections, values in parenthesis are the current settings of
our system.

13.3.2 Definitions

For clarity, we first define several terms. The first is the *harmonic intensity*
$E_t^{ch}(\omega)$ of the sound wave $x_t(\tau)$ at frame t. It is defined as

$$E_t^{ch}(\omega) = \sum_k \left\| H_{t,k}^{ch}(\omega) \right\|^2,$$

where $H_{t,k}^{ch}(\omega) = \sum_\tau x_t^{ch}(\tau) \cdot \exp(-jk\omega\tau),$

ch is a channel (left or right), τ is time, k is the harmonics index, $x_t^{ch}(\tau)$ is the
residual input, and $H_{t,k}^{ch}(\omega)$ is the sound component of the kth overtone. These
values are defined for binaural channels, that is, for the left and right channels.
Because some sound components are destroyed by interfering sounds, the
simple $H_{t,k}^{ch}(\omega)$ can not be used. To improve the sound consistency check,
valid overtones are used as segregation keys by each agent. An overtone of a
harmonics is *valid* if the overtone satisfies the harmonics consistency and
direction consistency. To improve evaluation of harmonic intensity, we also use
valid harmonic intensity, $\sum_{ch} \overline{E}_t^{ch}(\omega)$, which is defined as the sum of the
$\left\| H_{t,k}^{ch}(\omega) \right\|$ of valid overtones.

Harmonics Consistency. We define an overtone in a channel (left or right) as satisfying the harmonic consistency if the intensity of the overtone is larger than a threshold (Equation 13.2), and if the local time transition of the intensity can be approximated in a linear fashion (Equation 13.2) (Nakatani, et al., 1995a).

$$\left\| H_{t,k}^{ch}(\omega) \right\| > c \cdot \left\| \mathrm{DFT}_t^{ch}(k \cdot \omega) \right\|, (c = 0.15), \qquad \text{EQ. 12.1}$$

$$\sigma_{t,k}^{ch2} < p \cdot M_{t,k}^{ch}, (p = 0.05). \qquad \text{EQ. 12.2}$$

where $\mathrm{DFT}_t^{ch}(\omega)$ is the spectrum intensity of the actual input sound at frequency ω, frame $t,$, $\sigma_{t,k}^{ch2}$ and $M_{t,k}^{ch}$ are the following calculated values: each agent estimates the frequency $\overline{\omega}_{\tau,k}$ of the kth overtone at frame $\tau(= t,...,t+9)$ by tracking spectrum peaks of $\mathrm{DFT}_t^{ch}(\omega)$. Then, $M_{t,k}^{ch}$ is calculated as the mean value of $\left\| \overline{H}_{\tau,k}^{ch}(\overline{\omega}_{\tau,k}) \right\|, (\tau = t,...,t+9)$, and $\sigma_{t,k}^{ch2}$ is calculated as the variance of $\left\| \overline{H}_{\tau+1,k}^{ch}(\overline{\omega}_{\tau+1,k}) - \overline{H}_{\tau,k}^{ch}(\overline{\omega}_{\tau,k}) \right\|$, $(\tau = t,...,t+m-2)$, where $\overline{H}_{\tau,k}^{ch}(\overline{\omega}_{\tau,k}) = \left\| \mathrm{DFT}_t^{ch}(k \cdot \overline{\omega}_{\tau,k}) \right\|$.

Direction Consistency. Let *interaural phase difference*, $\Delta\omega_{t,k}$, be the phase difference of an overtone between left and right channels, that is,

$$\Delta\omega_{t,k} = \left\| \arg(H_{t,k}^{left}(\omega)) - \arg(H_{t,k}^{right}(\omega)) \right\|,$$

where $\arg()$ is a function returning an argument of a complex number. Let *interaural intensity difference* (IID), $\Delta I_{t,k}$, be intensity difference of an overtone between left and right channels, that is,

$$\Delta I_{t,k} = \ln(\left\| H_{t,k}^{left}(\omega) \right\|) - \ln(\left\| H_{t,k}^{right}(\omega) \right\|).$$

Let ω be frequency of an overtone. Then, an overtone satisfies the direction consistency with the interaural time difference (ITD), Δt, if there exists an integer n satisfying Equation 13.3,

$$(\Delta t - \theta_1) \cdot \omega \le \Delta\omega_{tk} + 2n\pi \le (\Delta t + \theta_1) \cdot \omega \qquad \text{EQ. 13.3}$$

and if Equation 13.4 holds.

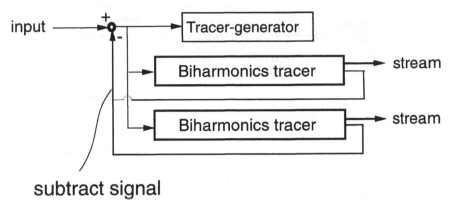

subtract signal

FIG. 13.4. Structure of BiHBSS.

$$\begin{cases} \Delta I_{t,k} > 0, & \text{if } \Delta t > 2\theta_1, \\ \Delta I_{t,k} < 0, & \text{if } \Delta t < 2\theta_1, \\ \theta_2 > \Delta I_{t,k} > -\theta_2, & \text{otherwise}, \end{cases} \qquad \text{EQ.13.4}$$

where $\theta_1 (= 0.08\text{ms}), \theta_2 (= 0.4)$ are constant threshold values. (Although the ITD is suitable for localization in low frequency regions and the IID is suitable for that in high frequency regions, they are treated uniformly in the current implementation.)

13.3.3 BiHBSS Agency

Figure 13.4 shows the structure of the BiHBSS agency. The BiHBSS agency segregates stream fragments according to the fundamental frequency, intensities and phase values of harmonics, and direction. There are two kinds of agents, a biharmonics tracer-generator and biharmonics tracers. The biharmonics tracer-generator finds new streams and generates biharmonics tracers. Biharmonics tracers tracks the streams.

Compared with the HBSS agency, the BiHBSS agency is extended to cope with binaural input and to permit biharmonics tracers to extract stream direction information and use direction consistency for segregating stream fragments. Each agent consists of a pair of agents that extract harmonic structure and another agent that coordinates interaural information (Figures 13.5 and 13.6). A harmonics extracting agent works for a channel input similarly to that of the HBSS agency, and the coordinating agent operate interaural information like direction.

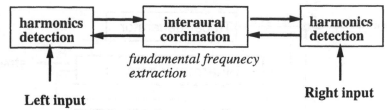

FIG. 13.5. Structure of a bi-watcher.

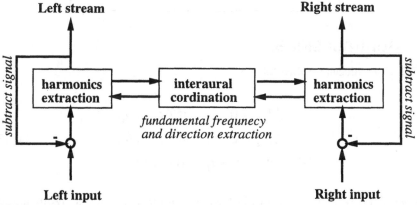

FIG. 13.6. Structure of a bi-harmonics tracer.

Biharmonics Tracer-generator. The biharmonics tracer-generator is designed as a set of biwatcher agents. A biwatcher consists of a pair of harmonics detection agents and an interaural coordination agent (Figure 13.5). A harmonics detection agent is similar to a pitch watcher in the HBSS. At each residual input, it detects a new harmonics stream whose fundamental frequency is in its own frequency region (about 25 Hz in width). The interaural coordination agent extracts fundamental frequency and it forces both harmonics detection agents to have the same fundamental frequency. A biwatcher is activated when the following conditions are satisfied at least in one of the left or right channel:

- $\overline{E}_t^{ch}(\omega) / E_t^{ch}(\omega) > r(r = 0.1)$, and

- there is a power peak near frequency ω in the residual signal.

Here ω is the frequency that maximizes $E_t^{ch}(\omega)$ within the region. Of all the active biwatchers, the biwatcher that gives the maximum $\sum_{ch} E_t^{ch}(\omega)$ generates a new tracer.

Biharmonics Tracer.   A biharmonics tracer consists of a pair of harmonics extraction agents and an interaural coordination agent (Figure 13.6).   A harmonics extraction agent is similar to a harmonics tracer in the HBSS.   The interaural coordination agent forces both harmonics tracer to have the same fundamental frequency and operates direction information.

A biharmonics tracer gets the initial fundamental frequency from a biwatcher when it is generated.   At each residual input, each biharmonics tracer extracts the fundamental frequency that maximizes the valid harmonic intensity, $\sum_{ch} \overline{E}_t^{ch}(\omega)$.   Each biharmonics tracer then calculates the intensity and the phase of each overtone in both channels by evaluating the absolute value and the phase of $H_{t,k}^{ch}(\omega)$.   Biharmonics tracers create subtract signals in a waveform by adjusting the phase of its overtones to the phase of the next input frame and recovers its signal by adding its subtract signal to the residual signal before calculating the fundamental frequency.   If there are no longer valid overtones, or if the intensity of the fundamental overtone drops below a threshold value, it terminates itself.

Each biharmonics tracer also extracts the direction information of a segregated stream fragment according to ITD, and IID of individual overtones at each time frame.   To calculate the direction, a biharmonics tracer uses a *direction histogram*.   A direction histogram is a vector value, each element of which represents a ITD candidate of stream direction.   The element containing the maximum value in the direction histogram decides the stream direction at each time frame.   A biharmonics tracer first checks Equations 3 and 4 for each overtone to each direction candidate, Δt.   If these equations hold for the k-th overtone to a direction candidate Δt^i, the biharmonics tracer add $\sum_{ch}\left\|H_{t,k}^{ch}(\omega)\right\|$ to the element of the direction histogram corresponding to Δt^i. Most harmonics coming from the same direction are usually added to the same element of the direction histogram, and, as a result, the direction element of the stream fragment becomes the maximum value.   (We assume that the ITD of binaural sounds arising from the same direction is the same between frequency regions, although there is actually a slight difference (Bodden, 1993).)

The direction of a sound stream extracted from a sound mixture, is not always stable.   The direction is often mistaken because a biharmonics tracer is affected by the other sound stream fragments.   Therefore, the extracted direction is not always credible.   Until the direction of stream fragments becomes stable, harmonics tracers use only Equations 13.1 and 13.2 in determining the validity of overtones.   When extracted direction of a stream fragment does not change during a duration threshold (=0.075 sec), the direction becomes stable and it can be useful as sound stream segregation clue.   Once extracted direction is detected to be stable, both harmonics consistency (Equations 1 and 2) and direction consistency (Equations 13.3 and 13.4) are used in determining the validity of overtones.

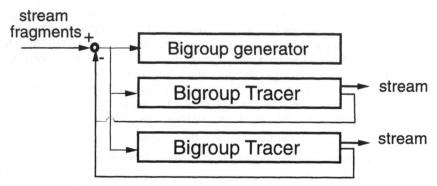

FIG. 13.7.   Structure of BiGrouping agency. BiGrouping agency segregates sound streams of individual sound sources by grouping stream fragments.

13.3.4   BiGrouping Agency

Stream fragments segregated by the BiHBSS agency are given to the BiGrouping agency. This agency finds streams, extracts streams by grouping stream fragments and monitors the conflict of stream fragment allocation to streams. Of course, these tasks are modeled by the RDA. Figure 13.7 shows the structure of the BiGrouping agency. It consists of a bigroup generator and bigroup tracers. When the bigroup generator receives a stream fragments, it detects a new sound stream and generates a new bigroup tracer. A bigroup tracer groups successive stream fragments according to their attributes. When two or more bigroup tracers conflict with each other at the same input stream fragment, the bigroup tracer tracing the sound stream with the closest attributes is selected to take the fragment. The residual fragment is only sent to the bigroup generator.

The BiGrouping agency uses fundamental frequency and direction information for grouping stream fragments. However, direction information of some stream fragments is not always extracted appropriately by the short-term analysis of the BiHBSS agency. The BiGrouping agency allocates such a non-directional stream fragment to a sound stream without direction consistency. And the direction is adjusted to that of the allocated sound stream of the fragment.

Bigroup Generator. After a stream fragment is put into the bigroup generator, the bigroup generator checks to see if the following conditions are satisfied,

(1)   the fragment exists longer than a time threshold (= 37.5 ms), and

(2)   no bigroup tracers pick up the fragment as its component.

If they are satisfied, the bigroup generator generates a new bigroup tracer.

Bigroup Tracer. A bigroup tracer gets an initial stream fragment from the bigroup generator when it is generated. The tracer then extracts a sound stream by grouping successive stream fragments segregated by the BiHBSS. The fundamental frequency and direction are decided from the attributes of one stream fragment, called a *central fragment*, among stream fragments composing its sound stream. First, the initial stream fragment is selected as the central fragment and the fundamental frequency of the fragment is used as the initial fundamental frequency. When direction information of a central fragment is first identified by the BiHBSS, initial direction is chosen as its direction. When one central fragment ends, another fragment is selected to be a new central fragment according to its attributes.

We use a measure of distance between attributes of streams and stream fragments to quantify the consistency of sound attributes. When a new stream fragment starts to be segregated by the BiHBSS and it becomes longer than a time threshold (= 37.5 msec), each bigroup tracer checks the distance conditions,

$$\left| F_f - F_s \right| < \theta_3, (\theta_3 = 300/600 \text{cent}) \qquad \text{EQ. 13.5}$$

$$\left| D_f - D_s \right| < \theta_4, (\theta_4 = 0.167 \text{msec}) \qquad \text{EQ. 13.6}$$

Here, F_f and F_s are fundamental frequency of the segregated stream fragment and that of the stream of the bigroup tracer. 1200 cent corresponds to an octave. D_f and D_s are direction of the segregated stream fragment and that of the stream of the bigroup tracer. If a fragment has bigroup tracers that satisfy these conditions during a specified period of time (= 0.15 sec), it gotten by the bigroup tracer whose sound stream has the attributes closest to those of the fragment according to the distance K,

$$K = \alpha \left| F_f - F_s \right| + (1-\alpha) \left| D_f - D_s \right|, (\alpha = 0.003), \qquad \text{EQ. 13.7}$$

(If direction information has not been useful yet, the second term of Equation 13.7 is ignored.) When the central fragment ends, another fragment that has the attributes closest to temporal attributes of the stream becomes the new central fragment. A bigroup tracer uses different values for θ_3 according to the situation. When two or more fragments exist at the same time, the bigroup tracer uses a smaller threshold value, that is, 300 cent, otherwise, it uses larger value, that is, 600 cent.

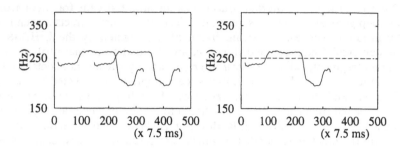

FIG. 13.8. Fundamental frequency patterns of input sound.

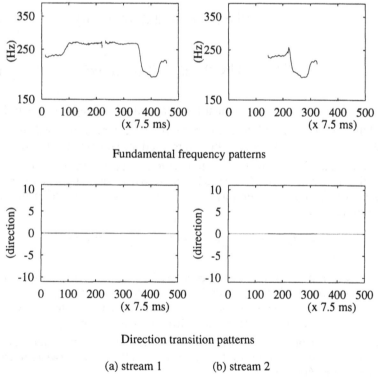

Fundamental frequency patterns

Direction transition patterns

(a) stream 1   (b) stream 2

FIG. 13.9. Sound streams segregated from mixed utterances of the sound set 1. Both utterances come from 90 degrees.

13.4 EXPERIMENTAL RESULTS

We evaluated the system by using two sound sets, sound set 1 and sound set 2, shown in Figure 13.8. Each sound set consists of two sounds. The sound set 1 has two utterances by a single woman. The second utterance is the same as the first one starting with a delay of 1 second. Its aim is to test binaural sound stream segregation performance of the system. The sound set 2 has the woman's voice and intermittent synthesized sounds (250 Hz) whose fundamental frequencies are close to each other. Synthesized sounds have overtones up to 6 kHz. Its aim is to test sound stream grouping performance of the system.

Sounds in a sound mixture are coming from directions within 0 and 180 degrees from the right-hand side. Direction of streams are extracted as an integer value between -10 and 10, where direction 9 corresponds to 0 degrees, direction 0 corresponds to 90 degrees (i.e., to the front), and Direction -9 corresponds to 180 degrees. The integer representing stream direction increases in proportion to the ITD.

Experiment 1. Utterances in the sound set 1 was put into the system. Each utterance in the sound mixture came from the same direction, 90 degrees. Segregated sound streams and their extracted direction are shown in Figure 13.9. The former part, ("ai"), of the second utterance and the latter part, ("eo"), of first utterance are segregated as the same stream fragment (stream 2). The other parts of two utterances are grouped together into stream 1. Direction of all streams is correctly detected direction 0 (90 degrees). Since two sounds have the same direction information, almost the same fundamental frequency and timbre, it is quite difficult to segregate these sound streams. Nevertheless, the harmonic structure of streams are segregated appropriately, and its direction is extracted correctly. However, sequential grouping of harmonic streams are not correct. Each utterance is not segregated as a stream and it is separated to be parts of different streams.

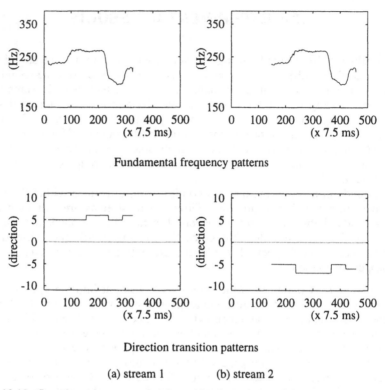

Fundamental frequency patterns

Direction transition patterns

(a) stream 1   (b) stream 2

FIG. 13.10. Sound streams segregated from mixed utterances of the sound set 1. The first utterance comes from 45 degrees and the second utterance comes from 135 degrees.

Experiment 2. The sound set 1 was used again. This time, two utterances were put into the system from different directions. The first utterance came from 45 degrees and the second utterance from 135 degrees. Segregated sound streams and their extracted direction are shown in Figure 13.10. Each utterance is segregated as a stream fragment. Each stream is almost correctly detected in direction 5 (45 degrees) or in direction -5 (135 degrees). When two utterances came from 0 and 180 degrees, almost the same results were obtained.

The results of Experiments 1 and 2 show that the system improves its segregation performance by using direction information.

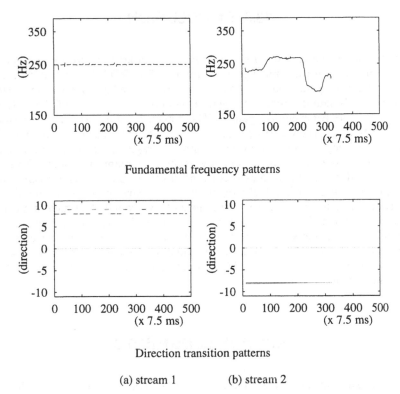

Fundamental frequency patterns

Direction transition patterns

(a) stream 1   (b) stream 2

FIG. 13.11. Sound streams segregated from mixed sounds of the sound set 2. The woman's voice comes from 180 degrees and the synthesized sounds come from 0 degrees.

Experiment 3. Sounds in the sound set 2 with various sets of directions were tested. When two sounds came from the same direction, 90 degrees, the woman's voice was segregated into three pieces of stream fragments and synthesized sounds were segregated pieces of stream fragments. But each woman's voice stream fragment was grouped with synthesized sound stream fragments, and the woman's voice can not be extracted as a stream. When the woman's voice is coming from 180 degrees and the synthesized sounds are coming from 0 degrees, the woman's voice and the intermittently occurring synthesized sound were segregated as streams, respectively (Figure 13.11). Direction of the woman's voice stream and direction of synthesized sound streams were approximately correctly detected to be direction -8 (170 degrees), and direction 8 (10 degrees). When two utterances came from 120 and 30 degrees, almost the same results were obtained. These results show that the system improves the sound stream grouping performance by using direction information.

13.5 CONCLUSION

The residue-driven architecture is a computational architecture for segregating sound streams with a multi-agent paradigm. The RDA is used to define an agency that segregates sound streams by tracking sound attributes. This chapter presented a binaural stream segregation system that consists of the BiHBSS agency and the BiGrouping agency. These agencies use harmonic consistency and direction consistency to segregate stream fragments and sound streams. The BiHBSS agency segregates harmonic stream fragments from binaural sound input, while the BiGrouping agency receives stream fragments from the BiHBSS and groups them to extract sound streams. Experimental results show that the system can effectively segregate sound streams using harmonic and direction consistency even from a mixture of two sounds that have almost the same fundamental frequency and the same timbre. The results also show that it can effectively group streams from a mixture of intermittent synthesized sounds and a woman's voice. We are now investigating the speech stream segregation agency and expect that it will enable computers to listen to several conversations at the same time.

ACKNOWLEDGMENTS

We thank Takeshi Kawabata, Makio Kashino, Hideki Kawahara, Rikio Onai and Ikuo Takeuchi for discussion on computational auditory scene analysis. We are also grateful to Toshio Irino and Tatsuya Hirahara for offering audio tools to collect binaural sound data.

REFERENCES

Bodden, M. (1993). Modeling human sound-source localization and the cocktail-party-effect. *acta acustica*, 1, 43-55.

Bregman, A. S. (1990). *Auditory Scene Analysis - The Perceptual Organization of Sound*. Cambridge, MA: MIT Press.

Brooks, R. A. (1986). A robust layered control system for a mobile robot. *IEEE Journal of Robotics and Automation*, RA-2 (1), 14-23.

Brown, G. J. (1992). *Computational Auditory Scene Analysis: A Representational Approach*. Ph.D. thesis, Univ. of Sheffield.

Cooke, M. P, Brown, G. J., Crawford, M., & Green, P. (1993). Computational auditory scene analysis: listening to several things at once. *Endeavour*, 17:4.

Costas, J. P. (1981). Residual signal analysis-a search and destroy approach to spectral analysis. *Proc. of 1st ASSP Workshop on Spectral Estimation*.

Kashino, K., Nakadai, K., Kinoshita, T., & Tanaka, H. (1995). Organization of hierarchical perceptual sounds: music scene analysis with autonomous processing modules and a quantitative information integration mechanism. *Proc. of IJCAI-95*.

Lesser, V., Nawab, S. H., Gallastegi, I., & Klassner, F. (1993). IPUS: an architecture for integrated signal processing and signal interpretation in complex environments. *Proc. of the 11th National Conference on Artificial Intelligence*, pp. 249-255.

Lyon, R. F. (1984). Computational models of neural auditory processing. *Proc. of ICASSP-84*.

Maes, P. (1991). Designing autonomous agents: theory and practice from biology to engineering and back. Special issue of *Robot and Autonomous Systems*. Cambridge, MA: The MIT Press/Elsevier.

Minsky, M. (1986). *Society of Minds*. New York: Simon & Schuster, Inc.

Nakatani, T., Okuno, H. G., & Kawabata, T. (1994a). Auditory stream segregation in auditory scene analysis with a multi-agent system. *Proc. of AAAI-94*.

Nakatani, T., Okuno, H. G., & Kawabata, T. (1994b). Unified architecture for auditory scene analysis and spoken language processing. *ICSLP-94*, 1403-1406.

Nakatani, T., Kawabata, T., & Okuno, H. G. (1995a). A computational model of sound stream segregation. *Proc. of ICASSP-95.*

Nakatani, T., Kawabata, T., & Okuno, H. G. (1995b). Residue-driven architecture for computational auditory scene analysis. *Proc. of IJCAI-95.*

Nawab, S. H. & Lesser, V. (1992). Integrated Processing and Understanding of Signals, 251-285. in Oppenheim, A.V. and Nawab, S. H. eds. *Symbolic and Knowledge-Based Signal Processing.* New Jersey: Prentice-Hall.

Okuno, H. G. (1993). Cognition model with multi-agent system (in Japanese), 213-225. in Ishida, T. ed. *Multi-Agent and Cooperative Computation II* (Selected Papers from MACC '92), Tokyo, Japan: Kindai-Kagaku-sha.

Ramalingam, C. S. & Kumaresan, R. (1994). Voiced-speech analysis based on the residual interfering signal canceler (RISC) algorithm. *Proc. of ICASSP-94.*

Schmids, R. O. (1986). Multiple emitter location and signal parameter estimation. *IEEE Trans. on Antennas and Propagation*, Vol.AP-34, No.3, pp. 276-280.

Slaney, M., Naar, D., & Lyon, R .F. (1994). Auditory model inversion for sound separation. *Proc. of ICASSP-94.*

Stadler, R. W. & Rabinowitz, W. M. (1993). On the potential of fixed arrays for hearing aids. *Journal of Acoustic Society of America,* 94 (3) Pt.1:1332-1342.

14

Discrepancy Directed Model Acquisition for Adaptive Perceptual Systems

Malini K. Bhandaru and Victor R. Lesser
University of Massachusetts at Amherst

For complex perceptual tasks that are characterized by object occlusion and nonstationarity, recognition systems with adaptive signal processing front-ends have been developed. These systems rely on handcrafted symbolic object models, which constitutes a knowledge acquisition bottleneck. We propose an approach to automate object model acquisition that relies on the detection and resolution of signal processing and interpretation discrepancies. The approach is applied to the task of acquiring acoustic-event models for the Sound Understanding Testbed (SUT).

14.1 INTRODUCTION

To meet the challenge of recognition in environments that are characterized by varying signal-to-noise ratio, unpredictable object activity, and possible object occlusion, *Adaptive Perceptual Systems* (Draper, Lesser 1993; Ming, 1990) have emerged. Recognition in such systems is dependent on the interaction between feature extraction and interpretation/matching: Failure to account for some or all data or adequately support a hypothesis triggers data reprocessing using alternate signal processing algorithms (SPAs) and/or parameters. Symbolic object models are used to interpret data, guide reprocessing and predict object interaction. Typically, these object models are handcrafted, a tedious and error prone activity that constitutes a knowledge acquisition bottleneck.

Model acquisition involves selecting for each object the appropriate SPAs (from a finite set of SPAs) and determining parameter settings for the selected SPAs such that the salient features of the object are extracted, to enable the induction of unambiguous object models. Automating model acquisition translates to automating the above search for SPAs and their parameterizations.

Our approach relies on the very mechanisms that make a system adaptive, namely those that detect signal processing inadequacies and suggest alternate processing strategies. To counteract the reduced top-down guidance due to a lack of knowledge about all the object classes, the learning system uses a greater number of signal processing discrepancy detectors that rely on comparing processing results of two or more SPAs. Where available, *generic* model expectations may also be exploited. The learning system has an additional focus: combining the multiple "views" of a signal that are exposed during the search process to generate a composite, more complete representation of the object, which better meets the prediction needs of adaptive perceptual systems.

In section 14.2 we briefly discuss learning effort in the area of model acquisition. In section 14.3 salient aspects of the Sound Understanding Testbed are presented along with an example of a sound event model. We next discuss the ramifications of parameterized SPAs in section 14.4 and the classes of discrepancies and their diagnosis in section 14.5. In section 14.6 we present the learning algorithm and some examples. The status of the work and our conclusions are presented in sections 14.7 and 14.8 respectively.

14.2 RELATED WORK

Automating model acquisition for adaptive perceptual systems has received relatively limited attention. Mori et al. (1987) addressed learning to identify speaker-modes, viewing it as a planning task that may possibly require elaboration and/or refinement, which in turn translates to extracting new features. When the feature set is determined to be insufficient, the domain expert intervenes to suggest alternate/additional features. Vadala (1992) presented an approach for acquiring models of sounds in the context of the SUT (Lesser et al., 1993), an adaptive perceptual system for nonspeech sound recognition. The models are acquired through adaptive processing of the signals, however, the work is limited in that firstly user guidance is necessary to initialize key SPA parameters and secondly, the number and nature of the sounds being modeled must be specified a priori. To summarize, the above rely on human intervention to initialize critical parameters and/or suggest alternate features to extract when feature inadequacies are encountered.

In the domain of vision, Ming and Bhanu's (1990) system learns models for aircraft identification when presented two dimensional images. The features used in the individual concepts are subsets of a fixed set of features (or SPA parameterizations) determined to be sufficient for the class of objects. The system, however, is not truly adaptive because it assumes that the images are adequately segmented. Likewise, Murase and Nayar's (1993) work on learning object models for recognition and pose estimation, using principal component techniques, assumes the availability of adequately segmented images. More

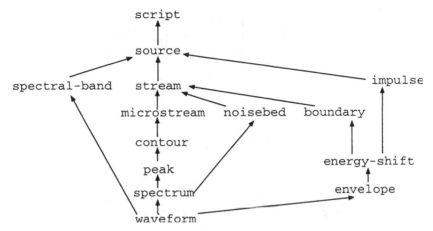

FIG. 14.1. Data Abstraction Levels used in the Sound Understanding Testbed.

recently, Bhanu et al. (1993) explored adaptive segmentation, although independent of model acquisition.

14.3 SUT: THE SOUND UNDERSTANDING TESTBED

The SUT seeks to identify acoustic-events given waveform data (a sequence of time-amplitude pairs). Signal understanding proceeds through a bidirectional search in the space of the SPAs and their parameters. The bottom up search is aimed at achieving signal processing results that are free of discrepancies (refer to section 14.5) that are detectable through the application of two or more SPAs of distinct strengths, and comparing their results. The top down search is guided by the desire to find valid signal interpretations based on expectations about the environment. SUT sound models are patterns of synchronized (Bregman,1990) sets of frequency components (Cohen, 1992).

Figure 14.1 shows the SUT data abstraction levels. Windowed waveform data that is analyzed for its spectral content is represented at the spectrum level. Peaks are localized regions of higher energy in a spectrum. Criteria such as the absolute cut off energy and the relative magnitude of a peak with respect to its neighbors determine what are considered as peaks in a spectrum. Contours are a sequence of peaks that move forward in time and share the same energy frequency trend. Noisebeds are regions of seemingly uncorrelated spectral activity. Contours that are consecutive in time and bear certain frequency and energy relationships are grouped together to form a microstream. Microstreams that are synchronized, either in their onset and/or offset times, energy behavior with respect to time, or whose frequencies are harmonically related, are grouped

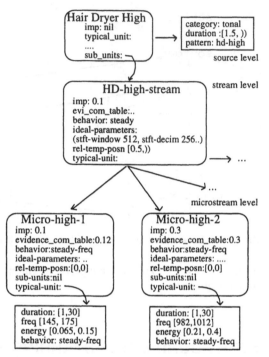

FIG. 14.2. SUT model for a hair dryer operating at high speed.

together to form streams. Groups of streams support a source level hypothesis. Periodic sources would display a repeating pattern of stream support units.

Figure 14.2 shows the SUT model for a hair dryer operating at high speed specified at the source, stream and microstream levels of data abstraction. The acoustic signal produced by a hair dryer is due to the working of a motor and the forcing of air through a nozzle. The component frequencies of the sound are harmonically related with a fundamental whose frequency is that of the power line. The relative energies of the harmonics are dependent on the speed of operation and the hair dryer construction (differing for different manufacturer models). Noisebeds, which are an artifact of the air flow through the nozzle of the hair dryer, surround the primary frequency components. The operation of a hair dryer exhibits two distinct phases, the transition or chirp phase that corresponds to the hair dryer being turned on or off, and a steady phase when it is operating at either high or low speed. The processing parameters that best bring out the time frequency characteristics in the two phases are in opposition (refer to section 14.4.1).

The source level unit in Figure 14.2 is made of a single stream level unit: hd-high-st, which in turn is made of several microstream units, two of which (μ-high-1 and μ-high-2) are shown. Note that the microstream durations are approximately equal and since their relative temporal offsets with respect to the stream start time are zero, their onset and end times are said to be in synchrony. A stream level representation that captures the complete behavior of a sound source, in terms of its constituent events, is possible. For example, the hair dryer sound may be specified as:

$$H\,Don\,(H\,Dhigh + H\,Dlow\,)^*\,H\,Doff$$

where *HDon, HDhigh, HDlow* and *HDoff* denote the hair dryer coming-on, operating at high speed, operating at low speed and going-off events, respectively. The above representation indicates that the *HDon* and *HDoff* events are mandatory, and that the *HDon* event precedes in time the HDoff event. In contrast, the *HDhigh* and *HDlow* events may each occur zero or more times (denoted by the *), and in any order (denoted by the +).

To date, the SUT database consists of 50 models. The models were acquired by manually analyzing several recordings of each sound. The tediousness of the task provides the motivation for this work. Our goal is to first acquire models for each of the constituent events of a sound source. Eventually, we plan to extend the work to building representations that capture complex temporal patterns.

14.4 RAMIFICATIONS OF PARAMETERIZED SPAS

In this section we discuss the effect of varying SPA parameters on the features extracted and introduce the notion of *SPA-correlate* to emphasize the connection between the two. We next discuss the inherent uncertainty in an object model, a consequence of its signal processing and interpretation history. The need for model synthesis is then discussed with a brief description of how it is achieved. Finally we discuss how generic models may be used to reduce search effort.

14.4.1 Parameterized SPAs

Distinct parameterizations of an SPA extract the same class of features, but the actual features extracted may be very different. This is because in the mathematical formulation of the SPAs, the parameters are used to capture assumptions about the underlying signal. To emphasize the relationship between the features extracted and its processing context (SPAs and their parameter settings), the features extracted are also known as *SPA-correlates*. Since some parameterizations of an SPA expose certain salient features while obscuring others, it is useful to compare the SPA-correlates obtained under different parameterizations of an SPA. This is possible using knowledge of the underlying signal processing theory which forms the basis of the SPA implementation (Lesser et al., 1993).

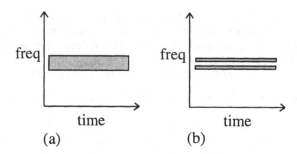

FIG. 14.3. Signal corresponding to two closely spaced steady frequency components, processed with (a) shorter and (b) longer STFT window. Note the better frequency resolution obtained in (b) due to the longer window.

For instance, consider the analysis of time amplitude waveform data corresponding to an acoustic-event composed of two constant frequency components with an inter-component spacing of 15 Hz. With a sampling frequency of 8 kHz, a Fourier Transform based algorithm for frequency analysis would be unable to expose the relevant frequency detail unless a data window that affords the minimum required frequency resolution is used. This is illustrated using the Short-Time Fourier Transform (STFT) algorithm (Nawab and Quatieri, 1988) for spectral analysis in Figure 14.3. Note that the uncertainty in frequency spread, that is, "width" of each component reduces when the data is processed with greater frequency resolution.

Second, certain SPA parameterizations afford a view of the signal data that leads to a *simple* physical explanation (Note: The principle also known as Occam's Razor, or the law of parsimony, may be stated as follows: Entities must not be multiplied beyond what is necessary; that is, an argument must be shaved down to its absolutely essential and simplest terms.). To illustrate this concept, consider the analysis of a near-linear rising chirp; a chirp in general is either a rising or falling frequency modulated component. Let the sound be sampled at a frequency of 8KHz. If the signal data is processed for spectral content using the STFT algorithm (Nawab & Quatieri, 1988) with a small window and narrow decimation (128 and 64 data points respectively) and then contouring applied, we would obtain results as shown in Figure 14.4a. Keeping all other processing parameters constant, but using a much longer STFT window (1024 data points), a broken curve as shown in Figure 14.4b would be obtained. Whereas the former may be interpreted as a "chirp", the latter could be interpreted as the presence of several sound sources, each of which emits a short burst of sinusoidal activity that is separated by approximately 10 Hz. Further, the latter interpretation indicates that the activity is highly synchronized in the sense that as a lower frequency source decays, the next higher one becomes active. Given the rarity of finding distinct physical events that are so highly synchronized, this interpretation requires too many assumptions making it not simple and hence discounted in favor of concluding that the signal was

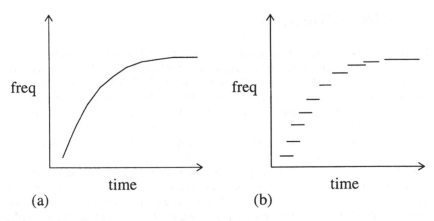

FIG. 14.4. Semi-linear chirp processed with (a) shorter and (b) longer STFT window. Note the broken contours of (b) due to insufficient time resolution.

inappropriately processed. The reasoning embodied in the learning system is that simple interpretations map to the notion of intrinsic object characteristics, originating in the physics of the excitation production mechanism.

14.4.2 Model Uncertainty

With any search process, there is inherent uncertainty due to the limited nature of our search. With DiMac, it is a consequence of the SPA parameterizations unexplored. The learning system maintains Symbolic Sources of Uncertainty (SOUs) (Carver, 1991) for each hypothesis to capture its uncertainty as a consequence of the uncertainty present in its support structures and the SPAs and parameters used in its creation. Symbolic SOUs recommend themselves because they may be examined to direct search in directions that promise to decrease uncertainty. Unacceptable levels of uncertainty in a model trigger further processing.

Before we describe in section 14.4.3 how the SOUs are used in model synthesis, we present a few examples of SOUs. At the contour level we have time and frequency uncertainty SOUs, in Figure 14.3 the uncertainty in frequency is represented schematically by the width of the contour. At the microstream level, separate SOUs are maintained for each microstream hypothesis to represent the uncertainty with respect to start and end time, steady energy and frequency. At the stream level, timing uncertainties such as that in the microstream are maintained along with an SOU that captures the likelihood of not having detected a component microstream. At the source level, we maintain SOUs with respect to the start and end times and uncertainty with respect to having missed a component stream.

14.4.3 Model Synthesis

In section 14.4.1 we illustrated how distinct SPA parameterizations are necessary to provide good time or frequency resolution. The dimensions of time and frequency are not orthogonal (Gabor, 1946) and SPAs cope differently: The STFT algorithm (Nawab & Quatieri, 1988) trades off resolution along the dimensions of time and frequency based on its parameter settings, whereas the Wigner algorithm (Claasen & Meclenbrauker, 1980) introduces cross terms, which often obscure the auto terms in multicomponent signals. Thus, in the case of sounds composed of both fine time and frequency detail, for example, a sound comprised of parallel chirps, it becomes necessary to combine "views" or SPA-correlates to generate a more complete representation. How are views synthesized? At the microstream level, when additional contour support becomes available, the time and frequency characteristics are updated based on whether the new support extends the microstream in time and whether it arises from a superior time or frequency context with respect to the earlier "best" support. The SOUs are accordingly updated. By selectively using the support data to update features, model synthesis is achieved and hypothesis' uncertainty with respect to time and frequency either decreased or maintained constant. Model synthesis is achieved likewise at the stream level; reprocessing at this level however concentrates on reducing uncertainty in its supporting microstreams.

14.4.4 Generic Models

For complex objects that exhibit many distinct high level features, the search effort involved may be significant. For example, the harmonic components of a hair dryer (any motor sound; refer to section 14.3) are tricky to identify without the use of a specialized SPA for harmonic enhancement. Such SPAs are however not routinely used due to their associated cost. In addition, for sounds that display varying stream level activity, such as in the case of a hair dryer, expectations that could be used to appropriately select SPAs and their parameter settings would help to reduce search effort. Generic models come into play to provide such support.

Generic Model Representation. How do we represent these models? Given that we seek models to represent a class of sounds, a representation as specific (absolute ranges for frequency, energy, and duration) as that in Figure 14.2 would be inappropriate. The model must capture only the intrinsic and not the incidental characteristics of the sound. To make this point clear: A motor sound should be recognized as such regardless of its intensity, the harmonics that may have been attenuated (a virtue of the physical casing of the motor), or the frequency of its fundamental (dependent on the power line frequency). Likewise a "chirp," or linearly modulated sinusoid, is a chirp regardless of its duration or its shrillness.

We seek to capture in a generic model a unique feature or a set of features that frequently co-occur and are representative of a sound class, such as motors, rings, buzzes, and impulsive activity. In fact, a hierarchy of generic models may be defined with leaf level models for simple high level properties such as tonal, impulsive, noise, and ringing. If we had a generic hair dryer model, it would be supported by the generic motor and noise models, with their interrelationships specified.

Indexing. The knowledge encapsulated in the generic models may be readily used if class information is provided along with the input signal. This, however, would be a step backward in the direction of automation. Instead, we favor using microstream and stream level characteristics to index into the generic model database. By operating at the contour or higher levels, we have data that is more appropriately processed due to the data data discrepancy detection and reprocessing that would have taken place. The goal here is to *recognize* the applicable generic models and is much like the SUT recognition system except for the different "object classes" and the different data used to index into the database. Applicable generic models can then be used to set up data expectation discrepancies and thus direct data processing. When evidence for the presence of lower level generic models accumulate, expectations for higher level models can be posted to recursively support search effort.

14.5 DISCREPANCIES AND DIAGNOSIS

In this section, we discuss discrepancy detection and diagnosis (Lesser et al., 1993), which form the backbone of our learning approach. Discrepancies fall into three categories: *data data*, *data expectation*, and *model model*. Their occurrence indicates inappropriate data processing. We discuss *diagnosis*, the process of explaining why a certain discrepancy may have occurred, with respect to the discrepancies. Occasionally two or more discrepancies are used in conjunction to arrive at a better diagnosis. The results of diagnosis are used to recommend alternate SPAs, parameters settings and a reprocessing plan that is most likely to eliminate one or more discrepancies.

14.5.1 Data Data Discrepancy

When the signal processing results obtained from the application of two or more distinct SPAs from a family of functionally similar SPAs are contradictory, we have a data data discrepancy. It indicates a need for SPA/parameter adaptation. For instance, Bitar et al. (1992) describe the use of the Pseudo-Wigner Distribution (Claasen and Meclenbrauker, 1980) in conjunction with the STFT

(Nawab and Quatieri, 1980) to detect inadequacies of time and frequency resolution.

14.5.2 Data Expectation Discrepancy

A data expectation discrepancy is encountered when signal processing results do not support expectations. The system is aware of object specific expectations when it encounters a new instance of an earlier encountered object or may have generic expectations if a generic object class has been identified. In addition, the system expects that all data must support a simple physical explanation (refer to section 14.4.1 for a detailed discussion).

A data expectation discrepancy is indicative of either invalid expectations or inappropriate data processing. If on data reprocessing using adapted SPAs and parameters, the discrepancy is resolved, it establishes that the data was originally inappropriately processed. For instance, when bottom up signal processing results in several "short contours" (refer to section 14.3); (note: Contour length categorization is a function of the number of supporting peaks, actual duration, and creation context which comprises of the spectral, peak, and contouring SPAs and parameters.) a discrepancy is flagged. It indicates that either our expectation that the source is tonal (as opposed to impulsive or noisy in nature) is false or that the data was processed with insufficient time resolution at the spectral level or that the contouring radii used were inappropriate.

When the diagnosis process encounters a data expectation discrepancy of type short contours, it checks whether there is in addition any data data discrepancy of type time resolution problem. If yes, it lends support to the hypothesis that the data may have been processed with insufficient time resolution. Its absence would support the theory that the contouring parameters were in error. The existence of an impulsive source would be discounted if the short contours extended over a tenth of a second of real time.

14.5.3 Model Model Discrepancy

When a newly generated model is ambiguous with respect to one or more earlier acquired object models, a model model discrepancy is flagged. It indicates that one or more object models require refinement or specialization through data reprocessing.

Diagnosis of a model model discrepancy involves examining the competing object models at successively lower levels of data abstraction in order to determine the cause of the discrepancy. This then literally translates to where in time frequency energy space and what manner of support evidence to seek in order to eliminate the discrepancy. Occasionally, such discrepancies cannot be removed if they originate chiefly due to an object model that was created earlier and whose signal data is not available for purposes of reprocessing. Under such circumstances, the database will be ambiguous until such time as a new instance of the object becomes available. Note that a fixed number (perhaps one) signal

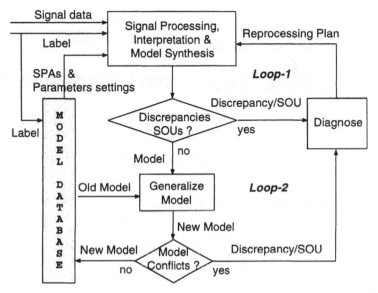

FIG. 14.5. Model acquisition algorithm.

data file could be maintained with each object encountered in an effort to mitigate order effects on database consistency.

14.6 DIMAC: DISCREPANCY DIRECTED MODEL ACQUISITION

In this section we present the learning paradigm and explain how it works through examples involving synthetic sounds.

14.6.1 Learning Algorithm

The learning task we seek to address is stated as follows: Given a set of training instances (signal/label pairs), and a finite set of parameterized SPAs, to seek for each training instance SPA parameterizations that serve to extract features that enable the induction of models that are collectively unambiguous and capture the intrinsic characteristics of the objects.

Loop-1 in Figure 14.5 seeks to resolve processing and interpretation discrepancies at successively higher levels of data abstraction. It involves processing data, checking for discrepancies, diagnosing the same, and reprocessing data after adapting SPAs and their parameters. Discrepancy resolution may entail reprocessing and interpretation at the current and one or

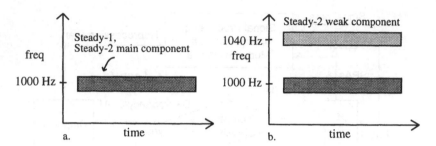

FIG. 14.6. Example: Discrepancy directed signal reprocessing for automated model acquisition.

more lower levels of data abstraction. Loop-2 terminates when all model model discrepancies are removed and model uncertainty is within tolerable bounds (refer to section 14.5.3 and section 14.4.2 respectively).

The algorithm is incremental in two respects, firstly objects may be trained for as and when they are encountered and secondly, the system incrementally refines object models when additional training instances of the same become available.

14.6.2 Examples

Two examples are presented to describe the sequence of events that would ensue during a learning session.

After encountering Example 1, one may wonder whether modeling would have been a one shot process if the best possible frequency resolution and the lowest possible energy threshold were used. The energy threshold is intentionally not maintained at the lowest level in order to minimize spurious effects, reduce noise and effectively model only the most essential of the components of a source. Example 2 illustrates a situation that shows working with the highest frequency resolution is not always the solution.

Example 1
Input: Two synthetically generated sounds: Steady-1 composed of a single frequency component at 1000 Hz and Steady-2 composed of a high energy component at 1000 Hz and a weak component at 1040 Hz, as shown in Figure. 14.6. Let both sounds be of equal duration.
Default Processing Context: STFT algorithm for spectral analysis with a window length of 1024 data points with data sampling at 10 kHz which provides a frequency resolution of 10 Hz. Peak selection based on accounting for a fixed percentage of total energy with an absolute energy threshold of 0.01 units of averaged energy. Contouring frequency radius of 10 Hz.
Assumption: The model database is initially empty. Once an object model has been incorporated into the database, its signal data is not available for purposes of reprocessing.

When the signal data for Steady-1 is analyzed using the default processing context, the object model generated indicates a single frequency component whose frequency lies in the range [995, 1005] Hz. No discrepancies are detected of a data data variety since time resolution problems are ruled out by the fact that the sound is steady and frequency resolution problems cannot arise because the sound indeed has only frequency component. Since the database is initially empty, as is to be expected, no model model discrepancies are detected. Note: Only relative energies of components are maintained in object models in order to generalize with respect to loudness.

When Steady-2 is encountered, once again no data data discrepancies are encountered. However, while seeking to include the model in the database a model model discrepancy is detected because the representations obtained for Steady-1 and Steady-2 are equivalent. The diagnosis process on comparing the models for the two sources and based on the fact that no frequency or time resolution problems were detected offers a single reprocessing strategy that suggests reducing the energy threshold in the hope of detecting another component. Let us say the absolute energy threshold is halved and now the weaker energy component is detected on data reprocessing. The model it gives rise to is distinct from the model for Steady-1 and is incorporated into the database. Had we first encountered Steady-2 as opposed to Steady-1, the reprocessing of Steady-1 data would not have yielded any disambiguating features. Consequently, the model, as is, for Steady-1 would have been incorporated into the database, resulting in an ambiguous database, and model refinement would be possible only when another instance of Steady-2 is encountered.

Example 2
Input: Consider the arrival of a Chirp as in Figure 14.4a.
Default Processing Context: as before.
Assumption: The model database contains models for Steady-1 and Steady-2.

The processing results obtained on using the default processing context for Chirp would be as shown in Figure 14.4b. If a model were generated based on the data, no model model discrepancies would have been detected since there is no ambiguity with respect to Steady-1 or Steady-2. However, at the contour level, a data expectation discrepancy would be flagged because the data would not support a *simple* explanation. In addition, a data data discrepancy would also be flagged by the STFT-Wigner discrepancy detection process (Bitar et al., 1992) indicating a time resolution problem. The diagnosis process on encountering these discrepancies would recommend data reprocessing using a shorter spectral analysis window and adapting the contouring parameters, that is, increasing the frequency and decreasing the time radii in order to follow more closely the transient nature of the signal. Finally, based on the smooth curve obtained, a model would be constructed and incorporated into the model database.

Recognition: Use of Object Models
Input: Waveform data comprising both Steady-1 and Steady-2 such that they start and end together.
Default Processing Context: same as before.
Assumption: Model database contains only the models for Steady-1, Steady-2, and Chirp.

We examine the sequence of events that would ensue during a SUT recognition run operating in Configuration II (Lesser et al., 1993). The spectral data obtained on STFT analysis is grouped over time into spectral activity bands, which are used to index into the model database. For the given scenario, spectral activity is restricted to a single band that indexes both Steady-1 and Steady-2. To disambiguate among the alternatives, the diagnosis component on examining the respective object models suggests data reprocessing using a reduced absolute energy threshold (based on the relative energies of the Steady-2 model) and paying close attention to the contour energies in order to establish whether Steady-1, or Steady-2 or both Steady-1 and Steady-2 are active.

Alternately, if the object model for Steady-1 and Steady-2 are the same because Steady-2 was encountered before Steady-1, that is, we have an inconsistent database, the recognition system would not be able to identify conclusively the scenario.

14.7 STATUS

The learning system produces models of high quality for real sounds that are purely tonal in nature or transient in nature. These correspond to signals that require one of good time or frequency resolution, with the system detecting which is necessary and to what degree. With the hair dryer coming-on sound, the system was able to detect that it needed high time resolution in the knee region and high frequency resolution in the plateau, but the model generated was unsatisfactory. This was because the learning system does not as yet handle long noisebeds, an intrinsic feature of sounds that have a forced air component. As a result, the contouring knowledge source was confused as to how to identify a trajectory in the sea of spectral activity in the knee region. Addressing this issue is our next step.

We will also be exploring the use of generic models to guide and hence reduce search effort. We will be exploring our intuitions regarding the identification of applicable generic models as well as investigating the reduction in computation obtained for different kinds of sounds.

Time savings are expected with any knowledge intensive approach as opposed to a pure search process. We will shortly be collecting timing data for model acquisition using both DiMac and a generalized time frequency, compute intensive approach (Jones & Parks, 1990) for determining the most appropriate processing parameters. We will also be comparing the models generated by the respective approaches to quantify the accuracy of our system.

14.8 CONCLUSION

The work presented indicates how the very ideas of adaptive signal processing can be used to address the knowledge acquisition bottleneck of acquiring object models. Our initial results with a limited number of discrepancy detection mechanisms have been promising. The success of the approach would ease the task of deploying recognition systems in new environments. A challenging next step would be to acquire object models without supervision as and when they are encountered (implies no object label) during a recognition session.

14.9 ACKNOWLEDGMENTS

The authors thank Frank Klassner and Hamid Nawab for valuable discussions.

REFERENCES

Bhandaru, M. K. & Lesser V. R. (1994). Automated object model acquisition for adaptive perceptual systems. Technical report CMPSCI TR94-24, Dept. of Computer Science, University of Massachusetts, Amherst, MA.

Bhanu, B., Lee, S., & Das, S. (1993). Adaptive image segmentation using multi-objective evaluation and hybrid search methods. *Working Notes of the AAAI Fall Symposium on Machine Learning in Computer Vision*, 30-34. Rayleigh, NC.

Bitar, N., Dorken, E., Paneras, D., & Nawab, H. (1992). Integration of STFT and Wigner analysis in a knowledge-based sound understanding system. *IEEE ICASSP '92 Proceedings*.

Bregman, A. (1990). *Auditory Scene Analysis: The Perceptual Organization of Sound*. Cambridge, MA: MIT Press.

Carver, N. (1991). A new framework for sensor interpretation: planning to resolve sources of uncertainty. *AAAI-91*.

Claasen, T. & Meclenbrauker, W. (1980). The Wigner distribution: a tool for time-frequency signal analysis. *Phillips J. Res.*, 35, 276-350.

Cohen, L. (1992). What is a multicomponent signal? *IEEE*.

Dawant, B. & Jansen, B. (1991). Coupling numerical and symbolic methods for signal interpretation. *IEEE Transactions on Systems, Man and Cybernetics*, 21(1).

Draper, B. (1993). Learning object recognition strategies. Dept. of Computer Science, University of Massachusetts, Amherst, MA. Available as technical report CMPSCI TR93-50.

Gabor, D. (1946). Theory of communication. *Journal of the Institute of Electrical Engineers*, 93, 429-441.

Lesser, V., Nawab, H., Bhandaru, M., Cvetanovic, Z. Dorken, E., Gallastegi, I., & Klassner, F. (1991). Integrated signal processing and signal understanding. Technical report CMPSCI TR91-34, Dept. of Computer Science, University of Massachusetts, Amherst, MA.

Lesser, V., Nawab, H., Gallastegi, I., & Klassner, F. (1993). IPUS: An architecture for integrated signal processing and signal interpretation in complex environments. *AAAI-93*, 249-255.

Ming, J., & Bhanu, B. (1990). A Multistrategy Learning Approach for Target Model Recognition, Acquisition, and Refinement. *Proc. of the DARPA Image Understanding Workshop*, 742-756. Pittsburgh, PA.

Mori, R. De., Lam, L., & Gilloux, M. (1987). Learning and plan refinement in a knowledge-based system for automatic speech recognition. *IEEE*, pp. 246-262.

Murase, H. & Nayar, S. (1993). Learning object models from appearance. *AAAI-93*, 836-843.

Nawab, H., & Quatieri, T. (1988). Short-time Fourier transform. *Advanced Topics in Signal Processing* Prentice-Hall, New Jersey.

Jones, D. & Parks, T. (1990). A high resolution data-adaptive time-frequency representation. *IEEE Transactions on Acoustics, Speech and Signal Processing*, 38 (12), 2127-2135.

Vadala, C. (1992). Gathering and evaluating evidence for sound producing events. Technical Report, Dept. of Biomedical Engineering, Boston University.

15

Auditory Scenes Analysis: Primary Segmentation and Feature Estimation

Alon Fishbach
Tel-Aviv University

This chapter describes a model of the primary stage of auditory scene analysis. This stage of analysis consists of two parts: segmenting the auditory scene to a collection of elementary auditory units and estimation of features such as onset, offset, and frequency and amplitude dynamics for each elementary unit. The formation of elementary units in the presented model is psychophysically and physiologically motivated. Both the segmentation and the feature estimation algorithms were tested on variety of auditory scenes and found to be a very good basis for computational auditory scene analysis grouping models.

15.1 INTRODUCTION

The acoustical environment around us is a rich source of information with respect to the location and characteristics of sound-producing objects. Auditory scene analysis is a process that is carried out by humans (and other organisms), that groups low level auditory entities to form higher level representations of the acoustical environment. Following this hierarchical description, auditory scene analysis grouping rules could be used to guide the construction of higher auditory forms from elementary units' building blocks. The grouping process is based on various acoustical properties and current auditory scene analysis models (Bregman, 1990) suggest that onset and offset time, temporal dynamics of amplitude and frequency, and spatial location are the most important features which govern auditory grouping decisions. However, there is currently no common definition of low level perceptual auditory elementary units and of the rules governing their formation.

In this chapter, we present an algorithm that can effectively segment a complex auditory signal into its elementary components, and estimate their onset/offset, amplitude and frequency temporal patterns. The elementary components formed by the algorithm along with their properties can serve as a basis for computational models of auditory grouping decisions.

In the following sections we briefly review the segmentation definition and algorithm and describe the procedure for estimation of elementary units' properties.

15.2 UNIT FORMATION

Bregman (1990) suggested that there is a unit-forming process that is sensitive to discontinuities in sound, but he gave no formal definition of these discontinuities. Some of the previous models do not define explicitly what an *elementary unit* is (Mellinger, 1991; Ellis, 1994). Others (Brown, 1992; Cooke, 1991) used algorithms to construct elementary units that are physiologically motivated but lack psychophysical support.

In this chapter, *elementary units* are defined as follows: First, units are defined by their borders, which is consistent with Bregman's assumption that the default tendency of our auditory perception is towards grouping. Second, borders are defined by discontinuities in intensity and phase, in time as well as across frequencies.

The effect of part of these discontinuities was tested psychophysically and physiologically and found to be consistent with the model. Olsen (Olsen, 1994; Olsen, 1995) found several neurons in the medial geniculate body of squirrel monkeys that are sensitive to intensity discontinuities in time when measured as the rate of intensity change. Very similar results were obtained psychophysically by Fishbach and Yeshurun (Fishbach, 1995) where subjects were asked to discriminate between continuous and discontinuous intensity change.

Short time discrete Fourier transform is used as the underlying representation in our model despite the fact that its characteristics do not match the filtering characteristics of the inner ear. We believe that at this stage of computational modeling of auditory processing it is better to use simple and tractable transformations as long as they do not prevent us from modeling the main grouping phenomena of auditory scene analysis.

Given this representation, intensity and phase discontinuities in time and across frequencies are mapped into large values of the directional derivatives of phase and absolute value of the complex spectrogram matrix (see Fishbach,

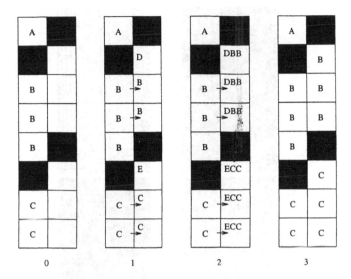

FIG. 15.1. Diagram of one recursive step of the segmentation algorithm: Each rectangle represents a spectrogram pixel; a black rectangle stands for boundary pixels. Letters A-E represent segment IDs. **0.** Beginning stage - leftmost column of the spectrogram is assigned with IDs. **1.** Each pixel assigns its ID to its right neighbor if possible. **2.** Trying to fuse as many pixels as possible in the current time-frame. **3.** Each pixel is assigned with the ID that appears most frequently.

1994) for a detailed mathematical treatment). Using the directional derivatives, we obtained four binary boundary maps; for each, a set pixel represents a border between auditory units. The boundary maps are merged into one boundary map that defines the unit's borders in the spectrogram. Parts (a) and (b) of Fig. 15.2 show a spectrogram of a male voice pronouncing the sentence "Why were you weary" overlaid by noise bursts (from Cooke, 1991) and its merged boundary map.

A recursive frame-by-frame segmentation algorithm is then applied, utilizing the boundary map to form groups of spectrogram's pixels, each such group representing an elementary auditory unit. The principle of the segmentation is to fuse together as many spectrogram pixels as possible while the boundary pixels mark the places in which fusion is impossible. Fig. 15.1 illustrates one recursive step of the algorithm. The segmentation algorithm resembles the birth and death algorithms used by Brown and McAulay and Quatieri (McAulay, 1986), but has the advantage of independence of horizontal spectral peak tracking, which allows formation of vertical segments. Part (c) of Fig. 15.2 shows the segmentation of the above example.

In order to evaluate the segmentation algorithm, we applied manual grouping operations to bind segments into groups. Then, we reconstructed each group by applying the inverse Discrete Fourier Transform selectively on that

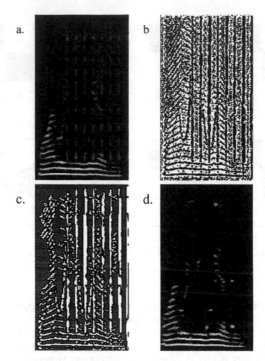

FIG. 15.2. Log log power-spectrogram of *"Why were you weary"* overlaid by noise bursts (from Cooke, 1991). (b) Binary boundary map of (a). (c) Segmentation of (a), each segment being represented by a white region enclosed by a black curve. (d) Log log power-spectrogram of the separated male voice.

group, and found that the groups formed indeed represent autonomic auditory events (Fishbach, 1994).   Part (d) of Fig. 15.2 shows the spectrogram of the separated male voice of the above example.

All the stages of the algorithm described are localized in time and frequency, and therefore can be parallelized. The non-parallel implementation of the model requires approximately 1.8 seconds of computation on a SGI Indigo R-4400 machine, for every second of raw input. These properties of the model make it suitable for real-time applications.

15.3 FEATURE ESTIMATION

Our current implementation handles only monaural information, so we describe methods of estimation of onset and offset timing and temporal dynamics of amplitude and frequency for each segment.

15.3.1 Onset and Offset

Onset and offset times are calculated according to the minimum and maximum time coordinates of the segment spectrogram's pixels. Converting time coordinates in the spectrogram domain (x), to time in the signal domain (T) is done as follows:

$$T = x\Delta t + W/2$$

where Δt is the time shift between successive windows (8 ms in our configuration) and W is the window length (64 ms). While enabling high frequency resolution, using such long windows yields a low time resolution. In our settings this means that the estimation of onset and offset time may deviate by around 4 ms. We show in section 15.3.5 how to refine this accuracy.

15.3.2 Frequency Estimation

The estimation of segment's frequency should be a vector of values which represents the "instantaneous" frequencies of the segment along time. In this process we utilize the phase of the segment's complex values. As shown in (Fishbach, 1994), for a spectrogram $Sp(.,.)$ of a sinusoidal signal with frequency F there exists:

$$\left\|\Phi(Sp(t,f)) - \Phi(Sp(t-1,f))\right\|_{2\pi} = \left\|2\pi\Delta tF\right\|_{2\pi}$$

$$\text{where } \|\Psi\|_{2\pi} = \begin{cases} -\Psi & \text{if } \Psi < 0 \\ 2\pi - \Psi & \text{otherwise} \end{cases}$$

where $\Phi(.)$ is the phase of a complex number, Δt is the time shift between successive windows, and f is the frequency coordinate which most closely matches the sinusoidal frequency. Because the normalized phase difference ranges between 0 and 2π, the resulting frequency ambiguity should be resolved using the f coordinate of $Sp(t,f)$. This calculation is carried out for each of the segment's pixels which share the same time coordinate, and then is averaged across them. If the standard deviation of all these frequency estimations is bigger than some threshold, then the frequency at that point is undecidable. The standard deviation threshold is adaptive and is adjusted according to the estimated frequency time derivative at that point. Segments with undecidable frequency normally represent abrupt events, usually heard as clicks.

The frequency estimation is very accurate for time-varying sinusoids (see Fig. 15.3). The modulation frequency for FM sinusoids can be assessed precisely as long as its time period is longer than half the spectrogram's window size. Currently this implies that the maximal frequency modulation rate that can

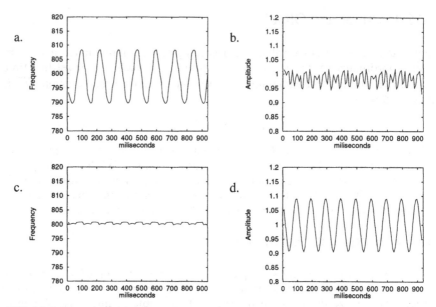

FIG. 15. 3. Frequency (a) and amplitude (b) estimation of FM modulated 800 Hz sinusoid with 8 Hz modulation rate and 8% modulation depth. Frequency (c) and amplitude (d) estimation of AM modulated 800 Hz sinusoid with 10 Hz modulation rate and 10% modulation depth.

be accurately estimated is 30 Hz. For FM rates smaller than 15 Hz, the FM depth estimation is accurate and starts to degrade as the FM rate increases.

15.3.3 Amplitude Estimation

The instantaneous amplitude estimation is based on the absolute values of the segment's pixels which share the same DFT window. Examining the case of one DFT window of a time-varying sinusoid suggests that there is one main parameter that determines the spread of intensity in the DFT window around the central frequency. The larger the range of frequency changes of the sinusoid, the more spread the intensity across the elements of the window.

Amplitude estimation is done numerically using a two dimensional lookup table. The first dimension is frequency range of change (R), and the second is width around the central frequency in the DFT window (W). Each bin in the lookup table contains the sum of absolute values of W pixels around the central DFT pixel, obtained by a DFT transform of a time-varying sinusoid with linear frequency change of R and amplitude 1. The estimated amplitude is the ratio between the sum of absolute values of concurrent segment's pixels, and the appropriate lookup-table's value.

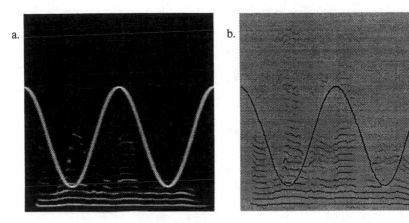

FIG. 15.4. a. Log-power spectrogram of *"Our lawyer will allow you a rule"* male voice accompanied by a siren, from (Cooke, 1991). b. Segments of (a); each curve represents the frequency estimation of a segment. The amplitude is represented by the brightness of the curve (on log scale).

As with frequency estimation, amplitude estimation is very accurate for time-varying sinusoids as long as the amplitude modulation rate is less than 30 Hz (see Fig. 15.3). The amplitude depth of amplitude modulated sinusoids is accurate as long as the modulation rate is smaller than 15 Hz and degrades as the rate increases.

15.3.4 Resynthesis

Although the feature estimation algorithms were found to be accurate by applying them on time-varying sinusoids with known parameters, we would like to verify how well they represent non-synthetic data by means of resynthesis. Each segment with decidable frequency is resynthesized as a time-varying sinusoid with instantaneous frequency and amplitude determined by linear interpolation of the segment's amplitude and frequency estimation. This method is very similar to that used by Cooke (Cooke, 1992). Several informal and non-comprehensive listening experiments on resynthesized speech and music show that the resynthesis is of very good quality.

15.3.5 Multi-resolution

Although the relatively low time-resolution of the model does not pose a real problem, a better resolution would certainly be a virtue. We wish to improve the time resolution of the model without degrading the frequency resolution. A general solution is not possible but under some restrictions a partial one may be obtained.

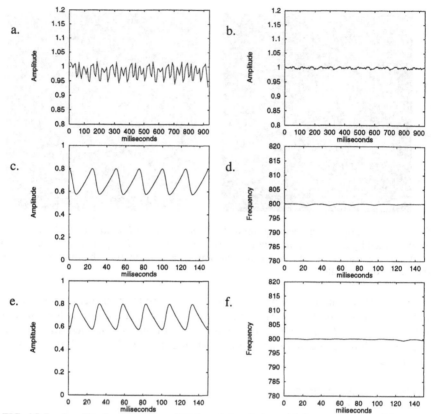

FIG. 15.5. Amplitude estimation of FM modulated sinusoid that was used in (a) and the same estimation using multi-resolution technique with 32 ms window length spectrogram (b). Amplitude (c) and frequency (d) estimation of exponentially ramped 800Hz sinusoid with half-life of 32 ms and repetition period of 25 ms which was used by Patterson (Patterson, 1994). Amplitude (e) and frequency (f) estimation of exponentially damped 800 Hz sinusoid with the same temporal parameters.

A second DFT spectrogram is used in addition to the segmented one, which is calculated using smaller time windows. Segments of the main spectrogram can be mapped to the second one, provided that the mappings do not intersect with any other segment's mapping. Calculation of onset and offset times is possible under less strict conditions, given that only the beginning and ending of the segment should fulfill the above requirement.

Up to now, the multi-resolution scheme was implemented only with segment's amplitude estimation. The results show improvement in amplitude estimation of AM sinusoids with high modulation rate, and in amplitude estimation of FM sinusoids as shown in Fig. 15.5.

15.4 RESULTS AND CONCLUSION

We presented a model of elementary units formation and feature estimation. The adequacy of the model's output in serving as input for grouping models was tested by two methods. The model was applied to a variety of synthetic time-varying sinusoids, and the features of the outcoming segments were compared to the original properties of the synthesized sinusoids. Additionally, the segmentation of auditory scenes which contain more than one source was tested by means of resynthesis of manually selected segment groups. Both methods suggest that our model can serve as a basis for computational auditory scene analysis grouping models.

The *elementary units* formed resemble Cooke's synchrony strands (Cooke 1992). The main difference is that in our work segment computations rely heavily on amplitude discontinuities in time and across frequencies, whereas synchrony strand formation does not utilize this information. Therefore it is reasonable to assume that the temporal aggregation of place groups into strands will not be consistent with the psychoacoustical results of experiments such as Bregman and Rousseau's (Rousseau 1991) and Fishbach and Yeshurun's (Fishbach, 1995)

The definition of *elementary units* in this model is psychoacoustical and physiologically motivated, although the current implementation of the model uses an underlying representation (DFT spectrogram) that is not physiologically compatible. One of our future research directions is to implement the above algorithms using representations that are derived from models of the auditory peripheral system.

15.5 ACKNOWLEDGMENTS

I would like to thank Prof. Y. Yeshurun for his continued guidance and support.

REFERENCES

Bregman, A. S. (1990). *Auditory Scene Analysis.*. Cambridge, MA: MIT Press.

Bregman, A. S. & Rousseau, L. (1991). Auditory intensity changes can cue perception of transformation, accompaniment, or replacement. Talk for the Psychonomic Society.

Brown, G. (1992). *Computational Auditory Scene Analysis: A Representational Approach*. Ph.D. thesis, Dept. of Computer Science, Sheffield University.

Cooke, M. (1991). *Modelling Auditory Processing and Organization*. Ph.D. thesis, Dept. of Computer Science, Sheffield University.

Cooke, M. (1992). An explicit time-frequency characterization of synchrony in an auditory model. *Computer Speech and Language*, 6:153-173.

Ellis, D. P. W. (1994). A computer implementation of psychoacoustic grouping rules. In *Proceedings of the 12th International Conference on Pattern Recognition*, volume 3, pp. 108-112, Jerusalem, Israel.

Fishbach, A. (1994). Primary segmentation of auditory scenes. In *Proceedings of the 12th International Conference on Pattern Recognition*, volume 3, pp. 113-117, Jerusalem, Israel.

Fishbach, A. & Yeshurun, Y. The role of intensity rate of change in auditory perception: discontinuity percept. In preparation.

McAulay, R. J. & Quatieri, T. F. (1986). Speech analysis/synthesis based on a sinusoidal representation. *IEEE Transactions on Acoustic, Speech and Signal Processing*, 34:744-754.

Mellinger, D. (1991). *Event Formation and Separation in Musical Sound*. Ph.D. thesis, Stanford University.

Olsen, J. Personal communication.

Olsen, J. (1994). Sensitivity of medial geniculate neurons in the squirrel monkey to rate of rise. In *J. Neurosci. Abstr.*, volume 20, p. 321.

Patterson, R. (1994). The sound of a sinusoid: Spectral models. *Journal of the Acoustic Society of America*, 96(3):1409-1418.

16

Cocktail Party Processors Based on Binaural Models

Jörn W. Grabke and Jens Blauert
Ruhr-University Bochum

Binaural cues are very important for the ability of humans to localize sound sources and to concentrate on objects like a desired speaker in a complex multispeaker situation. Therefore, modeling of the binaural capabilities enables algorithms to be developed that can be used to realize technical applications in the field of speech processing and enhancement systems.

A psychoacoustical motivated model is described, which has been developed in Bochum, Germany in recent years. The output of this model, the so-called binaural excitation patterns, include information about the spatial distribution of sound sources in the horizontal plane. The model was extended by units corresponding to higher stages of the auditory system, which analyzes the binaural patterns in order to separate a desired signal from a mixture of concurrent sounds.

To be used for technical applications, the binaural model has to be improved with respect to computing speed and performance in reverberant environments. New simplified algorithms to compute the binaural patterns, which are designed to run in real time on workstations, are presented. Technical applications of these algorithms are discussed.

16.1 INTRODUCTION

Binaural hearing offers a lot of advantages compared to monaural listening:

(1) sound sources can be localized in a three-dimensional space (spatial hearing),

(2) reverberation can be suppressed (precedence effect),

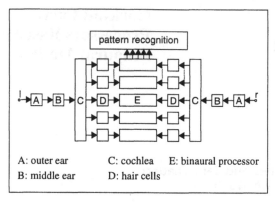

FIG. 16.1. Binaural model.

(3) interfering sound sources can be suppressed (cocktail party effect).

There have been various attempts in the past to develop models of binaural interaction which are able to reproduce these abilities. An overview of several types of models can be found in Stern (1988).

A binaural model to deal with the listed binaural phenomena must at least incorporate simulations of the external, middle, and inner ear, binaural processors to identify interaurally correlated contents of the signals of the cochleae and algorithms to analyze the information provided by the preceding blocks (Blauert, 1983). A model of binaural interaction developed at the Ruhr-University at Bochum which follows this general layout is discussed (see Figure 16.1).

16.2 STRUCTURE OF BINAURAL MODELS

16.2.1 External Ear

The external ears can be regarded as linear filters with transfer functions, which depend on the directions and distances of the sound sources. We model the external ears by using artificial head recordings as inputs to the following blocks or by filtering input signals with head-related transfer functions.

16.2.2 Middle Ear

For the modeling of binaural interaction it seems to be sufficient to model the middle ear as a linear bandpass filter.

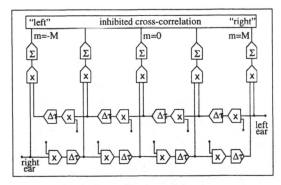

FIG. 16.2. Delay lines with inhibitory inputs.

16.2.3 Inner Ear

The inner ear performs a running frequency analysis of the incoming signals and an analog-digital conversion of the mechanical vibrations into firing patterns. A simple model for the frequency selectivity consists of a bank of adjacent bandpass filters each with a critical bandwidth. Converting the movement of the basilar membrane into spike series can be simulated by stochastic hair-cell models. The amount of computing required for these models is very high, so for practical applications we use a simplified model which computes firing probabilities by rectifying the bandpass signals and filtering the results with lowpass filters with cutoff frequencies of 800 Hz. The saturation of the firing probabilities of nerve fibers can be included by extracting the square root of the obtained time functions.

16.2.4 Binaural Processors

The binaural processor developed by Lindemann (1986) and Gaik (1993) is based on a running interaural cross-correlation, as suggested by Jeffres (1948). The cross-correlation is realized for each of 24 frequency bands so that the signals from the left and the right ear move in opposite directions along two delay lines and are multiplied at each tap. Additional connections in between the two delay lines implement a special inhibition mechanism, which is depicted in Figure 16.2.

This mechanism, named contralateral inhibition, shows important advantages compared to the simple model of Jeffres (1948):

(1) The periodicity of the cross-correlation function is suppressed and ambiguities are avoided,

(2) the contrast of the output patterns is enhanced, and

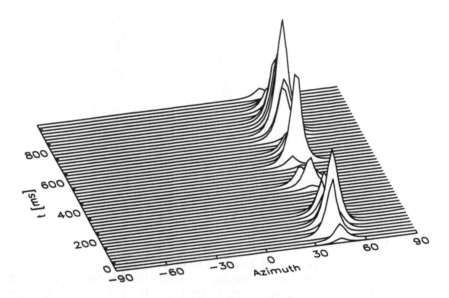

FIG. 16.3. Binaural excitation patterns of one source. The azimuth was 45 degrees.

(3) the binaural processor becomes sensitive to interaural intensity differences.

Monaural processors placed at the end of the delay lines facilitate the processing of monaural signals and signals from unnatural listening conditions.

The output of the binaural processor can be regarded as a simulation of neural excitation patterns. The analysis of these patterns gives information about the lateral distribution of hearing events and their psychoacoustical attributes. Figure 16.3 shows patterns for one speaker in non-reverberant environment.

16.2.5 Inhibitory Network

Wolf (1991) investigated the importance of onsets for localizing sound sources in reverberant environment. We used his results to implement a unit into the model, that improves the performance in these conditions (see Figure 16.4).

By connecting all the output units of the binaural processor with inhibitory paths it is possible to model the precedence effect (Grabke, 1994). The inhibition is adjusted in such a way that after a steep onset in the excitation patterns or any other cue representing a new hearing event in a plausible way, all the other output units are inhibited for some milliseconds. By doing so, further directional information from reflections is suppressed in accordance with the precedence effect.

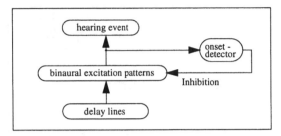

FIG. 16.4. Modeling the precedence effect.

16.2.6 Analysis Unit

This stage represents the more central functions of the auditory system with respect to binaural hearing. It analyzes the running binaural activity patterns in order to extract information about sound sources. The unit is able to predict the positions of hearing events in the horizontal plane and to discriminate between neural excitations due to several sound sources from different directions of incidence. The azimuth of sound sources can be determined from the peak positions in the binaural excitation patterns by performing the following steps.

First the correlation axis is transformed into an axis that directly represents the azimuth in the horizontal plane. Because the interaural time differences for one direction of incidence differ between the frequency groups, different transform functions for every frequency band are required. After this step a running average using a time constant from 1 ms to 100 ms is applied to smooth the resulting functions of time. Finally the excitations of all the critical bands are added to a weighted sum. The weighting can be interpreted as a spectral weighting of neural excitations. The weights for each band depend on the reliability of binaural information provided by the actual band. This is a preliminary realization of a cross-frequency-combination of binaural information.

The peaks in the total excitation pattern represent candidates for azimuths of hearing events. A candidate is accepted as a new hearing event, if the amplitude exceeds a given threshold.

16.2.7 Capabilities of the Binaural Model

The binaural model, as described above, is a model of sound-source localization that is able to recognize one or more sound sources. In doing so the following psychoacoustical effects can be reproduced: lateralization of one or more sound sources, summing localization, trading experiments with two resulting hearing events, and the precedence effect.

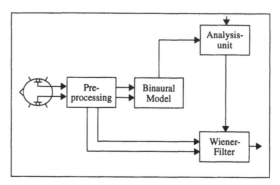

FIG. 16.5. Cocktail party processor.

16.3 APPLICATIONS OF BINAURAL MODELS

16.3.1 Cocktail Party Processing

Humans can concentrate on one talker in the presence of competing ones and noise from other sound sources. This ability is well known as the cocktail party effect. There have been several approaches to extract sound sources, especially of speakers from a mixture of acoustic signals, based on microphone arrays, adaptive filters, monaural cues, and binaural cues.

Slatky (1994) presented an algorithm which is able to separate two competing speakers by a mathematical analysis of the binaural cross-correlation.

The binaural model of Lindemann and Gaik was primarily developed to simulate of psychoacoustical effects, its structure is similar to that of the human subcortical auditory system. Since the output signals contain the information about the spatial distribution of sound sources and their specific energy, it is obviously possible to use the model for separating multiple sources. Based on this idea, Bodden (1993) developed a cocktail party processor which is shown in Figure 16.5.

First the position of the desired speaker is computed by the pattern recognizer. Then its signal-to-noise ratio (SNR) with respect to the competing speaker and noise signals is estimated for each frequency band. The energy of the speaker is calculated by windowing the binaural excitation with a window centered at the azimuth belonging to the speaker and summing up the resulting excitation. The ratio of this sum and the sum of the excitation all over the azimuth is a suitable estimation for the SNR. The SNR's of the critical bands are then used as weighting factors for a Wiener filter. The output signal of the cocktail party processor is the sum of the weighted critical band signals derived from the preprocessing filter bank. Using this a directional filter is realized with only two microphones.

In the case of two competing speakers the desired signal can be recovered to a degree of reasonable intelligibility even when its level is up to 15 dB lower than the other one. Bodden also performed an intelligibility test with hearing impaired subjects that showed a significantly higher comprehensibility in a cocktail-party situation if the speech signals were preprocessed by his system. The cocktail party processor thus proved to be suitable as part of a sophisticated hearing-aid.

16.3.2 Speech Recognition

Another application for binaural models is preprocessing speech signals to be fed into speech recognition systems. Bodden and Anderson (1994) tested the enhancement of phoneme recognition accuracy with binaural representations of signals in noisy situations compared to the monaural version.

They trained neural networks in binaural excitation patterns corresponding to the direction of incidence of the desired speaker. The results showed that both representations perform equivalently in clean conditions, but that the recognition performance of the monaural representation drops off as the SNR is reduced. The binaural representation provides an advantage of up to 20 dB SNR over the monaural one.

16.4 BINAURAL MODELING FROM THE VIEWPOINT OF ENGINEERING

Applications of the developed binaural algorithms are, for example, meters for psychoacoustical parameters, hearing aids, intelligent microphones and preprocessors for speech recognition systems. It is obvious that it is necessary to adjust the models' structure to fit these different tasks and to reach an optimum in precision and speed.

One problem concerning the realization of hearing aids using the described algorithms for cocktail party processing is the required computational time. The largest amount of time is required by the algorithm analyzing the interaural time and level differences. This is mainly necessary to reproduce laboratory results of psychoacoustical experiments, so that it does not seem to be essential for separating sources in normal conditions. To test this assumption, we developed a model with a simplified algorithm.

The structure of the Lindemann-Gaik model is reduced to kind of a coincidence detector. In order to decrease the necessary number of calculations only the peaks in the time signal are processed. After the preprocessing stages a peak recognizer extracts the information about the peaks and their amplitude and feeds it into the delay lines. Coincidence detectors between the delay taps

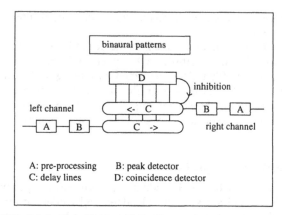

FIG. 16.6. Simplified model.

measure the time delays between two corresponding peaks and compute the energy of this coincidence. After a coincidence has occurred, both peaks are deleted from the delay line, which is equivalent to the contralateral inhibition in Lindemann's model. The structure is depicted in Figure 16.6.

By using this algorithm binaural excitation patterns similar to those of the original model can be computed (see Figure 16.7). Only small changes are necessary to the unit analyzing the binaural patterns and controlling the Wiener filter for speech enhancement.

From a mathematical point of view the modified algorithm can be interpreted as a running short-time crosscorrelation. Since it works in the time domain the time resolution is high and non-stationary signals can be analyzed exactly.

Only interaural time differences are measured so that level-based effects like stereophonis cannot be reproduced, however reproducing psychoacoustical effects is not necessary for technical applications as explained above. The new binaural model can easily be converted into fast computer programs. The computational effort has been reduced by a factor of five compared to the complete model, but the quality of the output signals of the cocktail-party processor is as least as good as before. At present we are trying to implement the simplified binaural model in real time on a standard workstation.

16.5 CONCLUSIONS

A simplified algorithm for analyzing binaural signals has been presented. The strategy was derived from a sophisticated binaural model, which is able to reproduce many psychoacoustical effects. The new models' structure has been

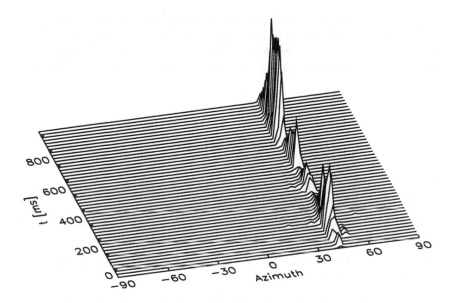

FIG. 16.7. Binaural excitation patterns of simplified model. Same input signals as in Figure 16.3.

adjusted to the task of enhancing speech from complex acoustic situations with low computational effort.

Previous investigations showed the benefit of binaural preprocessing for hearing-impaired persons (Bodden, 1993). A comparison with other speech enhancing systems like adaptive filters with respect to the profit in intelligibility is in preparation.

Another possible application is to develop intelligent microphones for things such as hand-free telephones. The binaural model is able to localize the speaker and the directional filter described above can be adjusted to the speakers' direction of incidence. The advantage of this system compared to conventional microphone arrays is that less space is needed for the micros and better performance is achieved in lower frequency regions.

In the moment the performance of the speech enhancement system is limited by the amount of reverberation in the input signals, which is a common problem most of these systems. The effect of strong reflections on the cocktail party processing is fluctuations in the amplitude envelope of the output signals. This can be avoided by using longer time constants in the processing unit, in turn lead to a lower directional filtering effect.

We expect further improvements in the cocktail party processor especially in the simulation of higher stages of the auditory system. One important task is the investigation of the mechanism combining information across frequency bands after the correlation process. Various experiments show the influence of common onsets in the time domain and influence of harmonic spectra in the frequency domain on building auditory objects (Bregman, 1990). We are now trying to use the binaural cues as an additional dimension, which could solve ambiguities occurring in the other dimensions such as crossing pitches of speakers. So binaural modeling also becomes very useful for auditory scene analysis.

REFERENCES

Blauert, J. (1983). *Spatial Hearing - The Psychophysics of Human Sound Localization.* Cambridge, MA: MIT Press.

Bodden, M. (1993). Modelling human sound source localization and the cocktail party effect. *Acta Acustica*, 1(1), 43-55.

Bodden, M. & Anderson, T. R. (1994). Improvement of speech recognition in noise. *WOS-report* 94-2094.

Bregman, A. S. (1990). *Auditory Scene Analysis.* Cambridge, MA: MIT Press.

Gaik, W. (1993). Combined evaluation of interaural time and intensity differences: Psychoacoustic results and computer modeling. *J. Acoust. Soc. Am.*, 94, 98-110.

Grabke, J. W. (1994). Modellierung des Praezendenz-Effektes [Modeling the precedence effect]. *Fortschritte der Akustik-DAGA'94*, DPG-GmbH, Bad Honnef, 1145-1148.

Jeffres, L. A. (1948). A place theory of sound localization. *J. Comp. Physiol. Psych.*, 61, 468-486.

Lindemann, W. (1986). Extension of a binaural cross-correlation model by contralateral inhibition; I. Simulation of lateralization of stationary signals. *J. Acoust. Soc. Am.*, 80, 1608-1622.

Slatky, H. (1994). *Algorithmen zur richtungsselektiven Verarbeitung von Schallsignalen - die Realisierung eines binauralen Cocktail-Party-Prozessor-Systems* [Algorithms for direction-selective processing of sound signals by means of a binaural cocktail party processor]. Duesseldorf: VDI Verlag.

Stern, R. M. (1988). An overview of models of binaural perception. National Research Council CHABA Symposium, Washington, DC.

Wolf, S. (1991). Untersuchungen zur Lokalisation von Schallquellen in geschlossenen Raeumen [Localization of sound sources in enclosed rooms]. Doctoral Dissertation, Ruhr-Universitaet Bochum.

APPENDIX

Results of binaural modeling

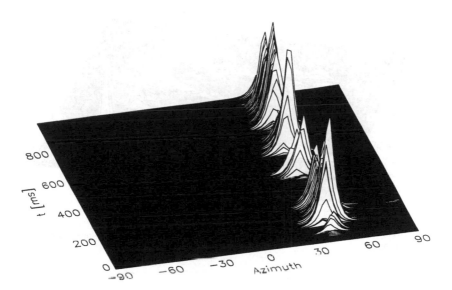

FIG. 16.8. One speaker in nonreverberant environment, original model.

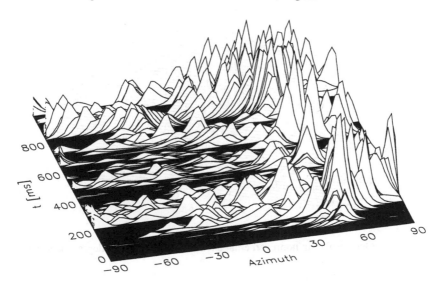

FIG. 16.9. One speaker in reverberant environment, original model.

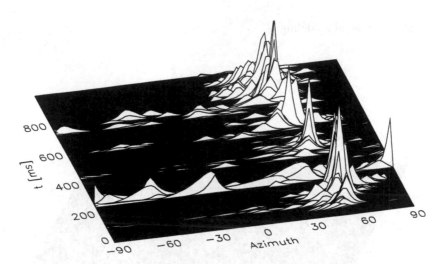

FIG. 16.10. One speaker in reverberant environment with modeling of precedence effect, original model.

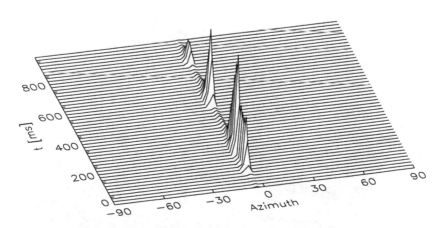

FIG. 16.11. One speaker in nonreverberant environment, azimuth 0 degree, new model.

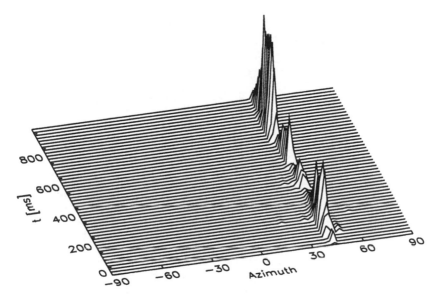

FIG. 16.12. One speaker in nonreverberant environment, azimuth 45 degree, new model.

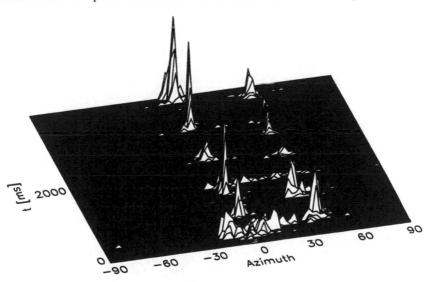

FIG. 16.13. Two speakers in nonreverberant environment, new model.

17

Midlevel Representations for Computational Auditory Scene Analysis: The Weft Element

Dan Ellis
International Computer Science Institute

David F. Rosenthal
Lernout and Hauspie Speech Products, Inc.

In this chapter we consider representations for use in models of the processing that occurs between the eardrum and the human conscious experience of sound. We first list 'good' properties for such midlevel representations, then present a framework within which to discuss some examples. We compare in detail two popular schemes - sinusoid tracks and correlograms - and propose a new representation, *wefts*, which seeks to combine their advantages.

17.1 INTRODUCTION

Midlevel representation is a term usually associated with computer vision, particularly the ideas of Marr (1982). It has since become accepted by many in the computer audition community (e.g. Bregman, 1990; Cooke, 1993; Brown, 1992) as an concept useful to models of hearing as well. Auditory perception may be viewed as a sequence of representations from 'low' to 'high,' where low-level representations are (roughly) those appropriate to describing the sound reaching the cochlea, and high-level representations are those to which we have cognitive access, such as "Mary telling John to buy bread," or "Bill playing the trombone with the TV in the background."

Between these two levels we presume there is a network of representations, which we label 'midlevel,' about which we have little direct knowledge; we are only beginning to understand the relevant physiology of the brain, and introspection is unlikely to be useful. Our knowledge of these representations arises chiefly from the constraints imposed on them at the lower and higher levels about which we have more information (Adelson, 1994).

Our view is that despite these challenges, we can understand midlevel representations by building computer models and that this is a fertile area for research in computer hearing. Success in such an endeavor will be rewarded by the construction of greatly improved robot perception systems, and a deepening of our understanding of perception in general.[1]

The relative wealth of knowledge about low-level hearing appears to have imposed an inordinately 'bottom-up' orientation on midlevel representations that have been proposed. While low-level processes are a useful source of interesting computations, it is important to retain a focus on the high-level constructs that may be useful for (computer or mammalian) hearing, and then to consider how these constructs may be computed.

17.1.1 Overview

In the next section we attempt to make a list of abstract qualities considered advantageous for a midlevel representation as we have defined it. Section 17.3 proposes three dimensions distinguishing different representations, and considers various examples from the literature in this light. Specifically, we consider the strengths and weaknesses of sinusoid tracks and correlograms, which leads us to propose a new representation, the weft, in section 17.4. Wefts address certain limitations of other these representations, as we illustrate with examples. We conclude in section 17.5 by restating our view of the role of representation in complete computational auditory scene analysis systems.

17.2 PROPERTIES DESIRABLE IN AUDITORY MIDLEVEL REPRESENTATIONS

What criteria should perceptual representations meet? The constraint from below on a midlevel hearing representation is that it may be computed efficiently from the input; from the upper side, the requirement is that it can readily answer the questions asked of it by the higher levels of processing (Winston, 1984). If we take the latter to include the full range of computer sound understanding applications, we can list the following desirable properties for midlevel representations:

Sound source separation. Arguably the *sine qua non* of hearing is the ability to organize sounds reliably according to their independent sources of production, roughly analogous to segmentation in vision. Natural sounds do not occur in isolation; they arrive at the ear from several sources as a complex mixture - meaning that the sounds may overlap in time, frequency, and other representational dimensions. To support a high-level description such as "Bill playing the trombone with the TV in the background," a midlevel representation should decompose sound to a granularity at least as fine as the sources of interest - in this case, pieces that can be labeled as TV noise or trombone.

Invertibility. We seek representations in which the original sound can be regenerated from its representation, although according to *perceptual* rather than exact criteria. That is, we want the regenerated sound be perceptually equivalent to the original, without requiring it to have an identical time-domain waveform. (Knight, 1994, suggested that exact regenerability may be viewed as a failure, because it shows that "nothing unimportant was discarded.")

More important is the separate *invertibility of meaningful parts*. By this we mean that the representation allows us to select a meaningful part of the sound - for example, the trombone without the TV noise in the example above - which can then be used to regenerate sound - in this case, a noise-free trombone.

We acknowledge that the human system does not actually resynthesize the sounds it represents internally, but the capacity for perceptual invertibility is effectively equivalent to a representation that captures *all* the relevant information. In addition, tractable inversion schemes will be important for applications of computational auditory scene analysis such as advanced hearing prostheses.

Component reduction. The initial sound may be regarded as a vast array of individual energy levels in time-frequency. As it is re-represented in successively refined ways, the number of objects in the representation should decrease, and the meaningfulness of each should increase. Note that this does not imply data compression; some of the midlevel representations discussed below require more bits to represent them than the original sound. What is important is that these representations group the elements of the original sound into a relatively small number of pieces, corresponding to meaningful structure in the original sound, and suitable for subsequent processing.

Abstract salience of attributes. The features made explicit by a representation should approach the perceptual attributes of our desired final result. In the interests of robust, modular development, we should define these features according to abstract source characteristics (onset of a new source) rather than specific algorithmic details (first difference of energy in each frequency band).

Physiological plausibility. Functional physiological knowledge becomes more and more scarce as we progress from the basilar membrane into the auditory cortex, but it still provides many interesting revelations. Because our goal is to understand and model the auditory system, we would be wise to respect this knowledge and not pursue hypotheses clearly inconsistent with physiology. This principle can, however, be difficult to interpret.

17.3 AN ANALYTIC FRAMEWORK FOR REPRESENTATIONS

So far we have considered criteria which affect the choice of representation; we turn now to actual candidates for hearing representations, and consider their various merits. We classify hearing representations according to three conceptual 'axes':

(1) the choice between fixed and variable bandwidth of the initial frequency analysis;

(2) discreteness, corresponding to the degree to which the representation is structured as meaningful chunks;

(3) the dimensionality of the transform - some representations possess a third dimension in addition to the usual pair of time and frequency.

By classification along these three axes, a given representation may be assigned a position on a cube, as in Figure 17.1. Several candidate hearing representations are now considered in relation to their positions.

In the rear lower corner is the Fast Fourier Transform (FFT). The FFT is computable by an efficient procedure, and its uses as an analysis tool (for instance, in the spectrogram) are familiar. It cannot, however, be considered an accurate model of the equivalent stage of the auditory system owing in part to the fixed bandwidth of its frequency bins. For a given FFT, increased resolution in frequency will come at the expense of resolution in time across all frequency channels, and vice versa. Physiological and psychological measurements of the auditory system indicate that it has the simultaneous ability to resolve frequencies and time variations in a manner not possible with a fixed-bandwidth model.

This limitation is addressed by moving along the "variable bandwidth" axis to the lower left-hand corner in our figure. This position is occupied by the constant-Q transform, implemented as a bank of filters whose bandwidths vary in proportion to their center frequency. This yields an analysis qualitatively similar to that performed by the cochlea. The constant-Q transform retains the simple mathematical formulation of the FFT, though its computational efficiency is not as great (Brown & Puckette, 1992).

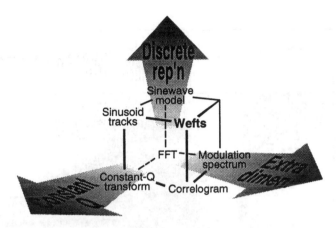

FIG. 17.1. Three dimensions for the properties of sound representations define a cube on which representations may be placed.

Both of these transforms are limited in representational power in that no higher level structure has been imposed; the signal is not 'chunked' into meaningful parts. In particular, our goal of separating the sound according to its sources requires additional processing.

The representations in the upper half of the cube, along the "discrete representation" axis, convert continuous transforms into discrete objects. We consider a particular example derived from the constant-Q transform, consisting of contiguous regions of local energy maxima in time-frequency called *sinusoid tracks* (Figure 17.2) (Ellis, 1992, 1994, based on the equivalent representation for FFT analysis introduced in McAulay & Quatieri, 1986). Tracks support the goals of component reduction and sound source separation, which is simplified to the problem of classifying a relatively small number of discrete objects - provided we are able to construct each track to represent energy from only a single source, which in practice can be very hard.

We can regenerate the original sound from its representation as tracks by using each track to drive a sine-wave oscillator. This technique can be applied to arbitrary subsets of tracks, addressing our criterion of "separate invertibility of meaningful parts." The resulting regenerated sounds possess a high degree of perceptual fidelity compared to the original, in spite of being poor approximations in a mean-squared error sense. In other words, they succeed in discarding unimportant information.

The third axis of Figure 17.1 is labeled "multiple dimensions." Representations on the right-hand face of the cube involve another dimension revealing extra information in addition to the time and frequency of a spectrogram. The interesting work of Kollmeier and Koch (1994) on the modulation spectrum falls into this category, although their goals are strictly practical (enhancement for hearing aids), and, hence, they are satisfied to use the FFT for their frequency analysis. For a more perceptually-motivated approach,

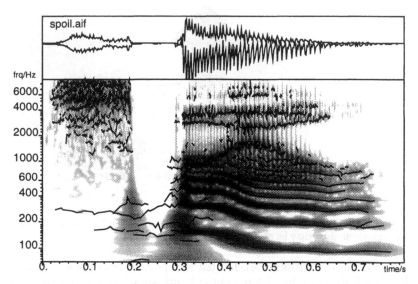

FIG. 17.2. A constant-Q analysis of the word "spoil" with the sinusoid track analysis overlaid. The upper panel shows the time waveform, and the lower panel shows time (left to right) versus log frequency (bottom to top); gray density shows the intensity of the constant-Q transform, and the black lines show the frequency contours of the sinusoidal tracks.

we consider here an example known as the *correlogram* (Duda, Lyon & Slaney, 1990; Slaney & Lyon, 1993), where the third dimension is the lag axis of short-time autocorrelations applied to the energy envelopes within each frequency band. Different parts of the spectrum whose intensity is modulated at the same rate will have similar profiles in this dimension, facilitating the goal of source separation. Note that the correlogram is normally calculated from the output of a cochlea model filterbank, although in principle it could be applied to any frequency decomposition. The computation of the correlogram is illustrated in Figure 17.3 and discussed in more detail in the following section.

17.3.1 Correlograms and Tracks

Why, then, is the extra dimension of correlograms useful? This is best seen by an analysis of a representational failing of tracks.

Periodic amplitude modulation is an important cue for sound source separation, since different parts of a sound sharing common modulation are most likely to have arisen from the same source. We would hope, then, that a sound representation would make such periodicity explicit. Unfortunately, the track representation may display periodicity in two separate ways. Consider a harmonic sound processed by a constant-Q filterbank (Figure 17.2): At the

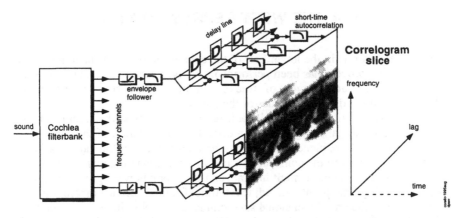

FIG. 17.3. The calculation of the correlogram. Sound is broken into frequency channels by the filterbank. The envelope of each channel is extracted by rectification and low-pass filtering. This envelope is multiplied by delayed versions of itself to detect correlation at various lags. The smoothed output of these multiplications forms a time-varying two-dimensional intensity display; dominant modulation periods appear as vertical structures in this view.

lower frequencies, the resolution is sufficient to separate the harmonics, which are analyzed to horizontal tracks whose periodicity is encoded in their frequency contours. At the higher frequencies, the bandwidth of the filters is broader and several adjacent harmonics will pass through the same filter. Beating between these harmonics will cause an amplitude-modulated filter output, resulting in the vertical stripes in the upper portion of Figure 17.2. Here, the signal's periodicity is primarily reflected in the *magnitude* variation of the tracks involved, not the frequency. This duality of periodicity encoding in the track representation means that subsequent processing must employ special strategies to group energy related by common period (Cooke, 1993; Ellis, 1994). (A fixed bandwidth analysis can avoid this problem but at an unacceptable cost in terms of overall time-frequency resolution).

By contrast, the correlogram effectively tags all of the energy related to the same amplitude modulation period with a common feature. If a particular band-pass filter output has a regular period of energy modulation, the autocorrelation of that channel will have an intensity peak at the lag matching that period. This scheme can be used to group all of the components - resolved and unresolved - resulting from a periodic sound, since they will all share this peak at the fundamental period.[2]

17.4 THE WEFT REPRESENTATION

The correlogram is in the lower half of the cube in Figure 17.1, meaning that no discretization process has been performed. We now turn to a discussion of how the structuring advantages of tracks can be combined with the greater dimensionality and improved cues of the correlogram.

The essence of the sinusoid track representation was the tracing of local maxima in the signal energy. Thus, a discrete representation of the correlogram volume (time × frequency × lag) can be constructed on the same basis. To construct tracks from the constant-Q transform, local maxima were picked in the spectrum at a particular instant in time, and these points were grown into tracks by connecting the points in adjacent time frames. In a correlogram, the equivalent of the one-dimensional spectrum is now a two-dimensional surface of frequency versus autocorrelation lag. We could pick the local maxima of this surface in two dimensions and follow them through time, forming a representation of one-dimensional contours extending, tendril-like, down the time axis of our three-dimensional volume.

However, the specific properties of the correlogram allow us to make a more useful representation that encodes common amplitude modulation across frequency bands directly into the basic elements. This provides a satisfying correspondence to the strong perceptual fusion of periodic signals revealed by introspection and experiments.

Thus our approach is to form elements that correspond to entire spectra defined by a common amplitude modulation. We do this by first picking peak lags for each frequency channel in our two-dimensional slice of the correlogram. This selects the dominant modulation periods in each band. We then look for particular periods that occur in a large number of frequency channels, by a mechanism similar to the 'summary autocorrelation' of Meddis and Hewitt (1991, 1992). The set of frequency bands that reflect modulation at a given period can then be tracked along time to form a kind of 'ragged ribbon' tracing the regions of spectral dominance of a particular modulated signal as it evolves in time, as illustrated in Figure 17.4. We call these structures *wefts*.[3]

The weft representation partitions the 3-dimensional correlogram space into chunks of sound characterized by a common period of amplitude modulation. Wefts appear to combine the important advantages of all the various representations we have discussed so far.

17.4.1 An implementation of wefts

We now describe our initial implementation of the weft representation. Figure 17.5 shows a block diagram of the system we have constructed. First, the sound (or sound mixture) is analyzed into frequency channels by a conventional constant-Q gammatone filterbank (Patterson, Holdsworth, Nimmo-Smith & Rice, 1988). This two-dimensional representation is then converted to a three-

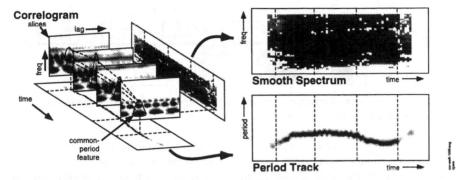

FIG. 17.4. Wefts are formed by making a vertical group of frequency bands that exhibit a common amplitude modulation peak in their autocorrelation lag axes at a given instant, then tracking this set of maxima through time. A weft is completely defined by its two projections, the period track and the smooth spectrum.

dimensional correlogram as described above by extracting the intensity envelope in each frequency channel (over a 2 ms window applied to the rectified signal), multiplying this envelope by delayed versions of itself, one for each sample of the lag dimension, then smoothing the output of this product over a 20 ms window, as illustrated in Figure 17.4.

The correlogram is a three-dimensional *volume* of intensity as a function of time, frequency and lag. This volume is then reduced to a two-dimensional summary autocorrelation (a function of time and lag) in the following manner: First, the peaks along the lag axis in the autocorrelation function are picked for each channel in each time frame. These autocorrelation maxima are 'spread' with a Gaussian window, then the peaks for all frequency channels at a given time frame are superimposed into a single function of lag showing the dominant periods present in the autocorrelations at that instant. Repeating this at each time frame gives the summary autocorrelation.

Since a periodicity of τ will result in peaks at lags of τ, 2τ etc. (and since resolved harmonics may contribute weak peaks at $\tau/2$, $\tau/3$ etc.), the peaks of the summary are searched to find a small number of fundamental periods that can explain all the lag-peaks. These periods are tracked through successive time-frames to produce separate 'period-tracks,' encoding the derived period of a detected periodic sound element as a function of time.

To recover the energy spectrum associated with a particular period track, the three-dimensional autocorrelogram is sampled at the time-lag coordinates corresponding to the period track. If a given frequency channel has an autocorrelation peak at that point, the square root of that intensity peak is copied to the 'smooth spectrum' corresponding to the period track, otherwise that time-frequency cell of the smooth spectrum is left blank (see Figure 17.3). This gives a complete weft, whose parameters comprise the period track (a function of time), and the smooth spectrum (showing the energy modulated at that period as a function of time and frequency).

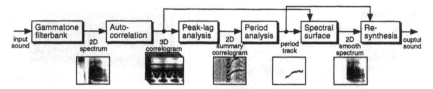

FIG. 17.5. A block diagram of our implementation of a weft analysis-synthesis system for separating mixtures of harmonic sounds.

Such a weft may be resynthesized in the following manner: A pulse train generated from the period track is filtered by a time-varying filterbank controlled by the smooth spectrum (which has been compensated for the spreading effect of the filterbank by a nonnegative least-squares approximation). This results in a resynthesis of the separated periodic signal based only on the information in the weft.

This algorithm draws heavily on previous autocorrelation-based schemes. The idea of sampling the autocorrelation at the lags indicated by the pitch track was suggested in Assman and Summerfield (1990), however, the idea of ignoring frequency channels that do not show a peak at or near that period is more similar to the system of Meddis and Hewitt (1992). Brown (1992) made unique periodicity assignments for each frequency channel which are biased by the dominant periods in the summary autocorrelation, largely equivalent to the strategy here of choosing dominant periods from the summary autocorrelation, then recruiting frequency channels to each. Brown's algorithm enforces exclusive allocation of a frequency channel to one period (as his resynthesis requires), whereas the system we describe can detect lag peaks for several different periods within a single channel; the value of this extra information remains to be quantified. The approach of extracting an entire pitch track for an identified sound object then using it to recover the spectrum, which dates back to Weintraub (1985), Stubbs and Summerfield (1988) and Quatieri and Danisewicz (1990) means that momentary distortions or uncertainties in the tracking can be overcome by interpolation from either side (which is indeed employed in the period-track extraction).

The most original aspect of the weft representation is the emphasis on resynthesis from the information in the representation alone, rather than by recourse to the original signal. The key advantage here is that a representation rich enough to reconstruct a whole signal is a fertile domain for signal modification and restoration in situations where more abstract constraints may be able to improve signal separation performance. A simple example of this is used in the resynthesis block, where small omissions in the recovered 'smooth spectrum' (i.e. time-frequency cells for which no corresponding autocorrelation peak could be located) can be simply interpolated from their neighbors, resulting in a more continuous resynthesis. A reconstruction algorithm that relies on unparameterized input for its fine detail will have difficulty in the case of this kind of corruption. A wealth of perceptual evidence suggests that high-level

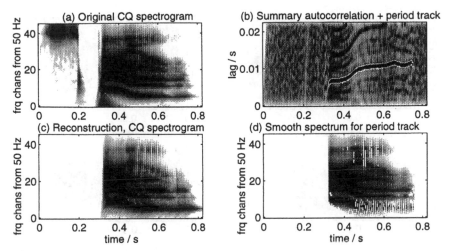

FIG. 17.6. Analysis of the single word "spoil". Panel (a) shows the constant-Q spectrogram of the original sound, analyzed by a 46-channel Gammatone filterbank with Q=8, 6 channels per octave, covering 50 Hz to 10 kHz. Panel (b) shows the summary autocorrelation for the entire sound. Note that the aperiodic fricative /s/ initial does not have any pronounced period. The period-track for the main weft is shown in outline. Panel (c) shows the constant-Q spectrogram for the resynthesis of that weft (i.e. the periodic portion of the original speech). Panel (d) shows the smoothed energy surface extracted from the correlogram volume for the period track indicated in (b); together, the period track and the smoothed surface completely specify the weft.

inference is central to the success of auditory perception (e.g. Warren, 1970; Ellis, 1993; Slaney, 1995).

Figures 17.6 and 17.7 give a graphical representation of the results of our process. Figure 17.6 considers the single word, "spoil," used in Figure 17.2 to illustrate sinusoid tracks. We see that the weft is able to model the periodic portion of the utterance, but not the aperiodic initial sibilant. Figure 17.7 deals with the analysis of a more interesting case of the mixture of two voices. In this case, the sound is the mixture of male and female speech used as example "v3n7" in Brown (1992). We see that two main pitch tracks have been formed over the summary autocorrelations, and there is a reasonable distinction between the spectrograms of the two reconstructed, isolated voices. These sound examples may be heard by visiting the World-Wide Web page for this chapter, URL: http://sound.media.mit.edu/~dpwe/research/pres/ijcai95.

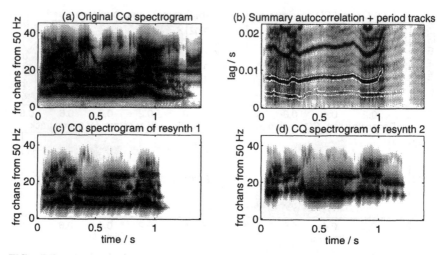

FIG. 17.7. Another example of weft analysis, for the mixture of male and female voices called "v3n7" in Brown (1992). (a) and (b) show the spectrogram of the original and its summary autocorrelation, as in figure 6. (c) and (d) show the constant-Q spectrograms of the two resynthesized voices, from the two wefts whose period tracks are outlined in (b).

17.5 SUMMARY AND CONCLUSIONS

Because computational models of auditory scene analysis constitute a relatively new endeavor, there is a temptation for each researcher to set out to solve the entire problem; in a sense, we do not understand the problem well enough to identify and disentangle more manageable mouthfuls. We should recognize that, like computer vision, there will be myriad different aspects to this area of research, and limiting focus to particular subproblems will probably prove rewarding. This chapter has attempted to move in that direction by reducing the scope to considering just the representation to be employed, its desirable properties, and what we can learn by analyzing the different representations that have been used in the past. Sadly, we have failed to reduce the problem by very much, since representation is intimately involved with aspects of scene analysis not addressed in this chapter, such as hierarchic abstraction and grouping rules.

Our introduction of wefts serves mainly to illustrate what we mean by a midlevel representation, and also follows our critique of sinusoidal tracks and autocorrelation to its logical next step. Certainly, our present system leaves much to be desired: At a detailed level, the extraction of the period track, and particularly the smooth spectrum, could benefit from a more careful analysis and implementation (for instance to establish the best way to estimate, from the autocorrelation, the magnitudes of two periodic energy bursts mixed in a single channel). More generally, a representation that relies on the concept of a period-track cannot handle the large aperiodic portion of our acoustic environment, usually categorized as noisy and/or impulsive. A hypothetical 'ultimate' midlevel representation will encompass these categories, although it is more likely that a variety of representations should be employed, each more or less appropriate for different kinds of sound.

In conclusion, we hope to have presented a useful theoretical framework, conceptual emphasis, and perhaps a practical tool for the representation of sounds in computational auditory scene analysis systems. Although our common goal of automatic models of the human perceptual system may still be some little way distant, we are at least on the path of what promises to be a fascinating journey.

17.6 ACKNOWLEDGMENTS

Thanks to Bill Gardner for specific comments on the chapter, and to all the members of the MIT Media Lab Machine Listening Group for providing such a swell working environment. This chapter would not have been possible without coffee.

REFERENCES

Adelson, E. (1994). Layered representations in vision. *Proceedings of the Abstract Perception Workshop*, Kanazawa, Japan. Tokyo: International Media Research Foundation. Available: WWW URL: http://dfr.www.media.mit.edu/people/dfr/apw–summary/apw-summary.html.

Assman, P. F. & Summerfield, Q. (1990). Modeling the perception of concurrent vowels: Vowels with different fundamental frequencies. *Journal of the Acoustical Society of America*, 88, 680-697.

Bregman, A. S. (1990). *Auditory Scene Analysis*. Cambridge, MA: MIT Press.

Brooks, R. A. (1991). Intelligence without reason. AI Lab memo 1293, Artificial Intelligence Laboratory, MIT. (Presented at the International Joint Conference on Artificial Intelligence, 1991.) Available: WWW URL: ftp://publications.ai.mit.edu/ai-publications/1000-1499/AIM-1293.ps.Z.

Brown, J. C. & Puckette, M. S. (1992). An efficient algorithm for computing the constant-Q transform. *Journal of the Acoustical Society of America*, 92, 2698-2701.

Brown, G. J. (1992). Computational auditory scene analysis: A representational approach. Ph.D. thesis CS-92-22, Department of Computer Science, University of Sheffield.

Cooke, M. P. (1993). *Modelling Auditory Processing and Organisation*. Cambridge, U. K.: Cambridge University Press.

Duda, R. O., Lyon, R. F. & Slaney, M. (1990). Correlograms and the separation of sounds. *Proceedings of the IEEE Conference on Signals, Systems and Computers*, Asilomar, CA.

Ellis, D. P. W. (1992). A perceptual representation of audio. MS thesis, Department of Electrical Engineering and Computer Science, MIT.

Ellis, D. P. W. (1993). Hierarchic models of hearing for sound separation and reconstruction. *Proceedings of the IEEE Workshop on Applications of Signal Processing to Audio and Acoustics*, Mohonk, NY. Available: WWW URL: ftp://sound.media.mit.edu/pub/Papers/dpwe-waspaa93.ps.gz.

Ellis, D. P. W. (1994). A computer model of psychoacoustic grouping rules. *Proceedings of the 12th International Conference on Pattern Recognition*, Jerusalem. Available: WWW URL: ftp://sound.media.mit.edu/pub/Papers/dpwe-ICPR94.ps.gz.

Knight, T. F. (1994). Lessons in perception from mammals. *Proceedings of the Abstract Perception Workshop*, Kanazawa, Japan. Tokyo: International

Media Research Foundation. Available: WWW URL: http://dfr.www.media.mit.edu/people/dfr/apw–summary/apw-summary.html.

Kollmeier, B. & Koch, R. (1994). Speech enhancement based on physiological and psychoacoustical models of modulation perception and binaural interaction. *Journal of the Acoustical Society of America*, 95, 1593-1602.

Marr, D. (1982). *Vision*. New York: W. H. Freeman.

McAulay, R. J. & Quatieri, T. F. (1986). Speech analysis/synthesis based on a sinusoidal representation. *IEEE Transactions on Acoustics, Speech and Signal Processing*, 34.

Meddis, R. & Hewitt, M. J. (1991). Virtual pitch and phase sensitivity of a computer model of the auditory periphery. I: Pitch identification. *Journal of the Acoustical Society of America*, 89, 2866-2882.

Meddis, R. & Hewitt, M. J. (1992). Modeling the identification of concurrent vowels with different fundamental frequencies. *Journal of the Acoustical Society of America*, 91, 233-245.

Patterson, R. D., Holdsworth, J., Nimmo-Smith, I. & Rice, P. (1988). Implementing a gammatone filterbank. APU report 2341, *Medical Research Council Applied Psychology Unit*, Cambridge, UK.

Quatieri, T. F. & Danisewicz, R. G. (1990). An approach to co-channel talker interference suppression using a sinusoidal model for speech. *IEEE Transactions on Acoustics, Speech and Signal Processing*, 38, 56-69.

Slaney, M. & Lyon, R. F. (1993). On the importance of time--a temporal representation of sound. In Cooke, M., Beet, S. & Crawford M. (eds.), Visual Representations of Speech Signals. New York: Wiley.

Slaney, M. (1995). A critique of pure audition. In Rosenthal, D. F. & Okuno, H. (eds.) Computational Auditory Scene Analysis. Mahwah NJ: Lawrence Erlbaum.

Stubbs, R. J. & Summerfield, Q. (1988). Evaluation of two voice-separation algorithms using normal-hearing and hearing-impaired listeners. *Journal of the Acoustical Society of America*, 84, 1236-1249.

Warren, R. M. (1970). Perceptual restoration of missing speech sounds. *Science*, 167, 392-393.

Weintraub, M. (1985). A theory and computational model of monaural auditory sound separation. Ph.D. dissertation, Stanford University.

Winston, P. H. (1984). *Artificial Intelligence* (2nd edition). New York: Addison Wesley.

NOTES

---

[1] We acknowledge that some researchers reject this symbolic, explicit approach to representation (e.g. Brooks, 1991); that debate is beyond the scope of this chapter.

[2] The fundamental harmonic would naturally be expected to have a peak at the fundamental period. Higher resolved harmonics will have additional peaks at shorter periods. To implement this scheme, the filtered signals should be half-wave rectified to ensure that resolved harmonics have amplitude modulation period corresponding to their frequency.

[3] From the American Heritage Dictionary: "The horizontal threads interlaced through the warp in a woven fabric," that is, a parallel set of threads.

18

The Complex-valued Continuous Wavelet Transform as a Preprocessor for Auditory Scene Analysis

Ludger Solbach and Rolf Wöhrmann
Technical University of Hamburg-Harburg

Jörg Kliewer
University of Kiel

In this chapter we draw links between the widely used gammatone filter auditory model and wavelet theory. From the viewpoint of wavelet theory the benefit from linking these research fields is a fast method for the computation of a time-scale representation. From the viewpoint of auditory filtering the benefits are the existence of methods for the detection of signal singularities and for resynthesis. Our method has proved to be useful for the analysis of music pieces with a limited spectral overlap of the different signal components. It has been implemented for further research in automated music transcription and auditory source separation, but might also be of interest for sound synthesis systems based on the analysis and transformation of acoustic signals.

18.1 LINEAR TIME-FREQUENCY DISTRIBUTIONS

Given a one-dimensional acoustic time signal s(t) a time-frequency distribution (TFD) is a two-dimensional representation of s(t) with time and frequency as its parameters. Basically a TFD tells us, which frequencies occur at which times in the input signal. There is, however, no unique TFD of a given signal s(t). For a

273

general overview of some different time-frequency representations the reading of (Hlawatsch & Boudreaux-Bartels, 1992) is recommended.

18.1.1 Why Wavelets?

The short-time Fourier transform (STFT) of a signal s(t) is given by

$$F_s(t,f) = \int_{-\infty}^{\infty} s(\tau) \cdot g^*(\tau - t) \cdot e^{-j2\pi ft} d\tau, \qquad \text{EQ. 18.1}$$

where the asterisk denotes complex conjugation. This is technically a windowed version of the Fourier transform, where a single window of constant shape is sliding along the time axis. The continuous wavelet transform (CWT) is given by

$$W_s(b,a) = \frac{1}{\sqrt{a}} \int_{-\infty}^{\infty} s(\tau) \cdot g^*\left(\frac{\tau - b}{a}\right) d\tau, \quad a > 0, \qquad \text{EQ. 18.2}$$

or equivalently in the frequency domain using Parseval's identity

$$W_S(b,a) = \sqrt{a} \int_{-\infty}^{\infty} S(f) \cdot G^*(af) e^{j2\pi fb} df, \quad a > 0, \qquad \text{EQ. 18.3}$$

As a is scaling the *mother-wavelet* g(t), it is called *scale parameter*, as b shifts it in time, we call it *shift parameter*. Given the Fourier transform G(f) of the mother wavelet g(t) the reconstruction of the time signal can be computed by

$$s(t) = c_g^{-1} \cdot \int_{-\infty}^{\infty}\int_{-\infty}^{\infty} W_s(b,a) \cdot \frac{1}{\sqrt{a}} g\left(\frac{t-b}{a}\right) da\,db, \qquad \text{EQ. 18.4}$$

where

$$c_g = \int_{-\infty}^{\infty} |g(t)|^2\, dt < \infty \qquad \text{EQ. 18.5}$$

The latter equation is a necessary and sufficient condition for admissibility of a time function g(t) as a mother wavelet. Equation 18.5 implies that

$$\int_{-\infty}^{\infty} |g(t)|^2 dt < \infty \qquad \text{EQ. 18.6}$$

and

$$G(0) = 0. \qquad \text{EQ. 18.7}$$

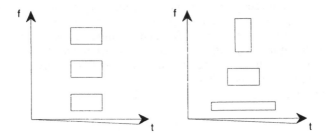

FIG. 18.1. Windowing in the time-frequency plane, left STFT, right CWT.

This means that admissible functions must have finite energy and their transfer functions must have at least one zero at $f = 0$. Functions satisfying these conditions look like short waves, which has been the reason for naming them *wavelets*.

In the past few years wavelet transforms have become an important tool for signal processing (Rioul & Vetterli, 1991). An important property of both the wavelet transform and the short-time Fourier transform is their linearity, which makes them more suitable for the analysis of multicomponent signals than quadratic TFDs suffering from cross-term artifacts.

The main difference between STFT and CWT lies in the variation of the analysis window along the frequency axis (see Figure 18.1). For both transforms the window size is constant. However, while the STFT window remains unaltered, the CWT window changes its shape due to the scaling factor a. In principle the CWT given by Equation 18.2 represents a constant Q filter bank with an infinite number of bands having infinitely small distances in frequency from the upper and lower neighbors. Because this idealized filter bank cannot be realized in practice, one has to modify Equation 18.2 by picking certain fixed values for a and b yielding a discrete approximation of the CWT.

In many applications wavelet transforms are applied to numerical and data compression applications employing orthogonal basis functions. These representations are redundancy free. Their drawback in signal analysis application lies in the fact, that they exhibit aliasing in the subbands corresponding to a lack of shift-invariance (Simoncelli, Freeman, Adelson, & Heeger, 1992). This means that if a signal is shifted in time, the wavelet coefficients might change drastically across scale instead of just being shifted in time as well. Thus, in an application like ours the continuous wavelet transform given by Equation 18.2 should be sampled on a fine grid, in order to maintain shift-invariance, yielding a *quasi-continuous* wavelet transform.

Contrary to the STFT the CWT realizes a logarithmic spacing of kernel functions in frequency space. The main advantage of logarithmic frequency spacing is that the trace of frequency strands obtained from the analysis of signals produced by weakly nonlinear systems (e.g. the harmonics) runs always parallel to the fundamental. This is an important advantage if we use the transform in pattern recognition applications and might also be the reason why the pitch perception in the human hearing mechanism exhibits logarithmic frequency resolution over a wide frequency range (Zwicker & Fastl, 1990). Because of the quasi-logarithmic organization of musical scales and of the frequency resolution in the human cochlea, the CWT is a more appropriate TFD of acoustic signals than the STFT.

The wavelets used throughout our work are analytic (progressive), that is they satisfy

$$\forall f < 0 : G(f) = 0$$

or at least close to being analytic. Thus, given the complex-valued filter outputs $y_i(t)$ the instantaneous frequencies

$$f_i(t) = \frac{1}{2\pi} \cdot \frac{d}{dt} \arg[y_i(t)]$$

can be estimated from the phases $\arg[y_i(t)]$ and the signal envelopes from the moduli $|y_i(t)|$, if the single signal components have a negligible overlap.

Practically the usability of the quasicontinuous CWT is affected by the following problems:

- Amount of generated data: The quasicontinuous wavelet transform yields a very redundant representation of the input signal. An easy way to reduce this redundancy would be to downsample each filter output to a rate of twice its bandwidth. However, we support the view that the reduction of redundancy should be performed in terms of detected features rather than by mere downsampling. Downsampling, even if maintaining reproducibility, has the tendency to obscure the features one wants to detect. For example, the estimation of the instantaneous frequency is negatively affected if the sampling is too sparse. Interpolation (that is, upsampling) would be required to increase precision.

- Computational burden: The computational burden of straight-forward finite impulse response (FIR) implementations can be significantly reduced by multirate techniques (Shensa, 1992), but the computing costs are still high. This is the reason why infinite impulse response (IIR) realizations like our gammatone approach are desirable.

18.1.2 Uncertainty

It is a well-known fact, that a function cannot be arbitrarily well concentrated in both time and frequency. The lower limit of the time-frequency window size is given by Heisenberg's uncertainty principle.

In order to define a measure for the resolution of a time-frequency representation, we have to define center and width of a signal. In the following we give definitions analogously to the corresponding definitions for mean and standard deviation in statistics. Be $s(t)$ a band-limited signal satisfying $\lim_{|t|\to\infty} s(t)\sqrt{t} = 0$. Given the energy of the signal

$$E = \int_{-\infty}^{+\infty} |s(t)|^2 \, dt = \int_{-\infty}^{+\infty} |S(f)|^2 \, df$$

we define the *window center* in the time domain as

$$t_0 = \frac{1}{E} \int_{-\infty}^{+\infty} t \cdot |s(t)|^2 \, dt. \qquad \text{EQ. 18.8}$$

In the frequency domain we have for the frequency window center

$$f_0 = \frac{1}{E} \int_{-\infty}^{+\infty} f \cdot |S(f)|^2 \, df. \qquad \text{EQ. 18.9}$$

The *time window width* be defined as

$$\Delta t_s = \sqrt{\frac{1}{E} \int_{-\infty}^{+\infty} (t-t_0)^2 \cdot |s(t)|^2 \, dt} \qquad \text{EQ. 18.10}$$

and the *frequency window width* as

$$\Delta f_s = \sqrt{\frac{1}{E} \int_{-\infty}^{+\infty} (f-f_0)^2 \cdot |S(f)|^2 \, df}. \qquad \text{EQ. 18.11}$$

One can derive for the area of the window in the time-frequency plane (Papoulis, 1987)

$$\Delta f_s \cdot \Delta t_s \geq \frac{1}{4\pi}, \qquad \text{EQ. 18.12}$$

meaning the localization of a signal in the time-frequency plane is principally affected by at least this uncertainty. The minimum area of $\frac{1}{4\pi}$ holds for the Gaussian function and its shifted variants in the time-frequency plane.

In terms of wavelet theory, however, these functions are not admissible, because Equation 18. 7 is not satisfied. Practically this can be neglected if the steepness of the filters is sufficiently high.

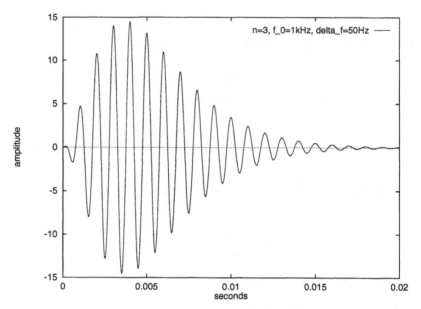

FIG. 18. 2. Gammatone filter of 3rd order, center frequency $f_0 = 1000$ Hz, bandwidth = 50 Hz.

18.2 THE GAMMATONE FILTER REALIZING A WAVELET TRANSFORM

As stated by Patterson (Patterson et al., 1992), the gammatone filter

$$g_r(t) = k \cdot \varepsilon(t)t^{n-1} \cdot e^{-\lambda t} \cdot \cos(2\pi f_0 t), n \geq 1 \qquad \text{EQ. 18.13}$$

where $\varepsilon(t)$ is the unit step function, n the filter order, $\lambda > 0$ the damping factor, k some normalization constant and f_0 the center frequency of the filter, can be a good approximation of the filtering in the human cochlea if the parameters are properly adjusted. See Figure 18.2 for an example of a gammatone impulse response. Different from Slaney's (Slaney, 1993) implementation we employed the analytic variant of the gammatone filter, constructed by applying the Hilbert-Transform

$$H(f) = -j \cdot \text{sgn}(f)$$

in the following way

$$g_{n,\lambda}(t) \quad = \quad g_r(t) + jH[g_r(t)]$$

$$= \quad k \cdot \varepsilon(t) \cdot t^{n-1} e^{-(\lambda - j2\pi f_0)t},$$

EQ. 18.14

with a Laplace transform of

$$G_{n,\lambda}(s) \quad = \quad k \cdot (-1)^{n-1} \cdot \frac{d^{n-1}}{ds^{n-1}} \cdot \frac{1}{s + (\lambda - j2\pi f_0)}$$

$$= \quad \frac{k \cdot (n-1)!}{(s + (\lambda - j2\pi f_0))^n}.$$

EQ. 18.15

For the scaled versions of $g_{n,\lambda}(t)$ we have

$$\frac{1}{\sqrt{a}} \cdot g_{n,\lambda}\left(\frac{t}{a}\right) \quad = \quad \frac{k}{\sqrt{a}} \cdot \varepsilon(t) \cdot \left(\frac{t}{a}\right)^{n-1} e^{-(\lambda - j2\pi f_0)t/a}$$

$$= \quad \frac{k}{a^{n-\frac{1}{2}}} \cdot \varepsilon(t) \cdot t^{n-1} e^{-\frac{1}{a}(\lambda - j2\pi f_0)t}.$$

It can be seen that $g_{n,\lambda}(t)$ is scaled by varying λ and f_0, while keeping their quotient constant. However, like the Gaussian the wavelet represented by $g_{n,\lambda}(t)$ is not admissible, as Equation 18. 7 is not satisfied. Again this fact can be neglected, if the damping of the DC component is sufficiently high.

18.2.1 Uncertainty of the Gammatone Filter

Without loss of generality we consider the continuous lowpass prototype of the gammatone filter, given by

$$g_{LP}(t) = k \cdot t^{n-1} e^{-\lambda t}$$

EQ. 18.16

with its Fourier transform

$$G_{LP}(f) \quad = \quad k \cdot \int_0^\infty t^{n-1} e^{-\lambda t} e^{-j2\pi ft} \, dt$$

$$= \quad \frac{k \cdot (n-1)!}{s^n}\Big|_{s=\lambda + j2\pi ft}$$

EQ. 18.17

$$= \quad \frac{k \cdot (n-1)!}{(\lambda + j2\pi f)^n}.$$

For the energy of the system's impulse response we find

$$E(n,\lambda) = \int_0^\infty (k \cdot t^{n-1} e^{-\lambda t})^2 \, dt$$

$$= k^2 \cdot (2\lambda)^{1-2n} \Gamma(2n-1)$$

EQ. 18.18

with the gamma function

$$\Gamma(x) = \int_0^\infty e^{-t} t^{x-1} \, dt, \text{ for } x > 0.$$

EQ. 18.19

For energy normalization we set

$$k = \frac{1}{\sqrt{(2\lambda)^{1-2n} \Gamma(2n-1)}}.$$

EQ. 18.20

From Equation 18. 8 and Equation 18. 9 follows

$$f_0 = 0,$$

EQ. 18.21

$$t_0(n,\lambda) = \frac{\Gamma(2n)}{2\lambda\Gamma(2n-1)} = \frac{2n-1}{2\lambda}.$$

EQ. 18.22

Using Equation 18.10 and Equation 18.11 we get

$$\Delta t(n,\lambda) = \frac{1}{2\lambda} \cdot \sqrt{2n-1}$$

$$\Delta f(n,\lambda) = \lambda \cdot \sqrt{\frac{2^{2n-5} \Gamma(n-1)\Gamma\left(n-\frac{3}{2}\right)}{\pi^{\frac{5}{2}} (n-1)^{-1} (2n-1)^{-1} \Gamma(2n)}}$$

EQ. 18.23

$$= \frac{\lambda}{2\pi} \cdot \sqrt{\frac{1}{2n-3}},$$

EQ. 18.24

with the use of Legendre's equation $\Gamma(x)\Gamma\left(x+\frac{1}{2}\right) = \frac{\sqrt{\pi}}{2^{2x-1}}\Gamma(2x)$. Finally, we find for the window area in the time-frequency plane

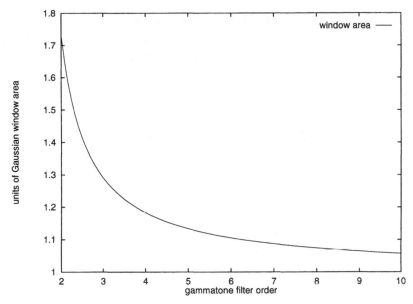

FIG. 18.3. Window area of the gammatone filter of order n, the unit of the y-axis is the area of the Gaussian function.

$$(\Delta f \Delta t)(n) = \frac{1}{4\pi}\sqrt{\frac{2n-1}{2n-3}}.$$

EQ. 18.25

As expected, the window size depends only on the filter order n. After fixing the order to a certain value the form of the window is determined by the parameter λ. For $n \to \infty$ the window size approaches $\dfrac{1}{4\pi}$ which is the Gaussian window size (see Figure 18.3). This result could be expected from the fact that an increase of n in Equation 18. 17 is basically (up to a scaling factor) equivalent to the repeated convolution of the first order gammatone lowpass prototype, which, due to the central limit theorem, is known to converge against the Gaussian function.

18.2.2 Group Delay of the Gammatone Filter

For the phase of the n-th order gammatone lowpass prototype we have

$$\arg[G_{n,\lambda}(f)] = -\arg([\lambda + j2\pi f]^n)$$

$$= -n \cdot \arctan \frac{2\pi f}{\lambda},$$

EQ. 18.26

so the group delay $\delta(f, n)$ is

$$\delta(f, n) = -\frac{1}{2\pi} \cdot \frac{d}{df} \arg[G_{n,\lambda}(f)]$$

$$= \frac{n \cdot \lambda}{\lambda^2 + 4\pi^2 f^2}.$$

EQ. 18.27

With Equation 18.24 we have for $f = 0$

$$\delta(0, n) = \frac{n}{\lambda} = \frac{n}{2 \cdot \pi \cdot \Delta f(n, \lambda) \cdot \sqrt{2n - 3}}.$$

EQ. 18.28

For Δf fixed, $\delta(0, n)$ reaches its minimum for $n = 3$, since

$$\frac{d}{dn} \cdot \frac{n}{\sqrt{2n - 3}} = \frac{n - 3}{(2n - 3)^{\frac{3}{2}}} = 0 \Rightarrow n = 3$$

and

$$\frac{d^2}{dn^2} \cdot \frac{n}{\sqrt{2n - 3}} = \frac{6 - n}{(2n - 3)^{\frac{5}{2}}} > 0 \text{ for } n = 3.$$

18.2.3 Parameterization of the Gammatone Filterbank

For implementation on a digital computer the transfer function of the continuous gammatone filter given by Equation 18.14 has to be transformed into a discrete equivalent. Cooke (Cooke, 1993) realizes his gammatone filter bank by shifting the input signal in frequency by multiplication with the phasor $e^{j2\pi f_0 kT}$, where f_0 is the center frequency of the band to be calculated, passing the output through the gammatone low-pass prototype and shifting the resulting signal back. The advantage of this method is the applicability of the impulse invariant transform for the discrete-time approximation of the low-pass prototype. This, however, is paid by an increase in computational burden. For this reason we decided to use true band-pass filters. Different from Slaney's (Slaney, 1993) implementation, we did not apply the impulse invariant transform. The results in favor of this transform in (Cooke, 1993) were given for the discrete-time approximation of base band filters.

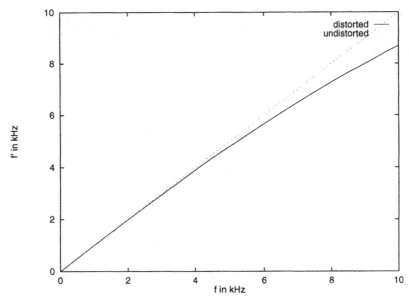

FIG. 18.4. Frequency distortion caused by the bilinear transform, sampling rate is 44.1 kHz.

For true bandpass filters, however, their transfer function having a nonzero value at half the sampling rate, we have to be aware of the fact, that we face an increased influence of aliasing components. This is why the bilinear transform, given by

$$s = 2f_s \cdot \frac{z-1}{z+1},$$ EQ. 18.29

where f_s is the sampling rate, is the most appropriate transform, because it avoids the appearance of aliasing components. An important point is that the application of the bilinear transform causes a distortion of the frequency axis, which can be calculated by inserting $z = e^{2\pi \frac{f'}{f_s}}$ and $s = j2\pi f$ into Equation 18.29, yielding

$$f' = \frac{f_s}{\pi} \cdot \arctan \frac{\pi \cdot f}{f_s}.$$ EQ. 18.30

For example, given a sampling rate of 44.1 kHz, if the center frequency f_0 of a continuous gammatone filter equals $f_0 = 10$ kHz, the center frequency of its discrete approximation moves to 8.688 kHz (see Figure 18.4). In order to account for this distortion when calculating the center frequencies of the filters we compute f_0 for which f_0' equals the desired frequency and use the result for the calculation of s_0, yielding

$$s_0 = \lambda - j2f_s \cdot \tan\frac{\pi \cdot f_0}{f_s} \qquad\qquad \text{EQ. 18.31}$$

instead of $\lambda - j2\pi f_0$. Now, inserting Equation 18.29 into Equation 18.15 we get

$$G_n(z) = \frac{k \cdot (n-1)!}{\left(2f_s \cdot \frac{z-1}{z+1} + s_0\right)^n} = \frac{k \cdot (n-1)! \cdot \alpha^n (z+1)^n}{(z+\beta)^n}$$

with

$$\alpha = \frac{1}{s_0 + 2f_s} \text{ and } \beta = \frac{s_0 - 2f_s}{s_0 + 2f_s}.$$

Finally, for the coefficients a_i and b_i of the n-th order discrete gammatone filter

$$G_n(z) = \frac{\sum_{i=0}^n a_i \cdot z^{-i}}{\sum_{i=0}^n b_i \cdot z^{-i}}$$

we find by simple calculation

$$a_i = k \cdot (n-1)! \cdot \alpha^n \cdot \binom{n}{i} \text{ and } b_i = \beta^i \cdot \binom{n}{i}, \qquad \text{EQ. 18.32}$$

with k given by Equation 18.20.

Of course, not only the center frequency is affected by the application of the bilinear transform. Another effect is that the transfer function of a bandpass filter is getting more and more asymmetric with increasing center frequency. This fact, however, does not necessarily have a negative effect on the quality of the preprocessing for sound analysis, since even the human cochlea seems to exhibit asymmetric filter slopes as physiological measurements indicate, see for example (Zwicker and Fastl, 1990).

18.3 DETECTION OF SINGULARITIES

We are interested in points of time, where something new happens. Thus two questions arise

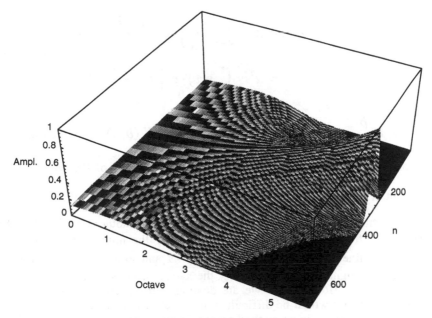

FIG. 18.5. Impulse analyzed by an analytic Gaussian wavelet.

(1) How can we see if there is "something new"?

(2) What is a "point of time" in the time-frequency plane?

Both questions can be answered by the notion of singularities. We call a distribution singular in a point x, if it cannot be represented in a neighborhood of x by a function being differentiable in that point. An example for such a point is $t = 0$ for the dirac $\delta(t)$, the unity step $\varepsilon(t)$ or the linear ramp $\varepsilon(t) \cdot t$.

Proposition: If the input function $s(t)$ satisfies a homogeneity condition of the form $s\left(\dfrac{t}{a}\right) = a^{-\gamma} s(t)$, with a real-valued and constant, we have

$$\frac{b}{a} = const \Rightarrow \arg(W_s(b,a)) = const.$$

Proof: From the definition of $W_s(b,a)$ in Equation 18.2 we have

$$W_s(b,a) = \frac{1}{\sqrt{a}} \int_{-\infty}^{\infty} s(\tau) \cdot g*\left(\frac{\tau - b}{a}\right) \ d\tau, \quad a > 0.$$

From the homogeneity of $s(t)$ follows

$$W_s(b,a) = a^{\gamma-\frac{1}{2}} \int_{-\infty}^{\infty} s\left(\frac{\tau}{a}\right) \cdot g^*\left(\frac{\tau-b}{a}\right) \quad dr$$

$$= a^{\gamma+\frac{1}{2}} \int_{-\infty}^{\infty} s(\tau) \cdot g^*\left(\tau-\frac{b}{a}\right) \quad dr.$$

EQ. 18.33

That is, if $\dfrac{b}{a}$ is constant, the wavelet transform $W_s(b,a)$ is constant up to a factor $a^{\gamma+\frac{1}{2}}$ for the modulus. In particular, the phase of the $W_s(b,a)$ is constant, if $\dfrac{b}{a}$ is constant. Q. E. D.

Note that this is independent of the choice of the mother wavelet. For the dirac $\delta(t)$ for example, we have $\gamma = -1$, for the unity step $\varepsilon(t)$ holds $\gamma = 0$, the linear ramp $\varepsilon(t) \cdot t$ features $\gamma = 1$ and so on. The estimation of the factor γ might be used to identify the nature of the singularity (Mallat & Zhong, 1992). However, as we are dealing with superpositions of several signals, the result would be difficult to interpret in most nontrivial cases. At least we can hope to find a method for the detection of the characteristic phase pattern to localize onsets of signals without using any specific knowledge of the signal characteristics.

From the above follows, that if such a singularity occurs, one can observe a characteristic local pattern in time-scale space formed by lines of constant phase spreading across all scales and converging to a single point for $a \to 0$. Each line l_i of this pattern satisfies $b = t_0 + \kappa_i \cdot a, \kappa_i \in IR$. As our scale axis is logarithmic we have $a' = \log a$ instead and thus $b = t_0 + \kappa_i \cdot e^{a'}$. So we explicitly know that the lines of constant phase caused by a singularity are exponentials in the time-scale plane. See Figure 18.5 for an example showing the impulse function analyzed by an analytic Gaussian wavelet. The zero phase is colored black, fading to white versus 2π.

The idea behind our singularity detection algorithm is to select a line of the phase pattern and set up a matched filter for it. The modulus of the filter output in response to the singularities to be detected should be as high as possible at the selected line of the pattern. If the modulus is small, the line can be distorted more easily by other superimposed signals. For the Gaussian the zero phase line at $t = 0$ is a promising candidate, whereas the same line cannot be used for the gammatone filter, since its modulus is zero. For the gammatone filter we choose the line where the impulse response reaches its maximum. For finding the maxima of the impulse responses across scales we calculate the first derivative of Equation 18.16:

$$\frac{d}{dt}\left(t^{n-1}e^{-\lambda t}\right) = e^{-\lambda t}t^{n-2}[n-1-\lambda t], \text{ for } t > 0. \qquad \text{EQ. 18.34}$$

Thus we have

$$\frac{d}{dt}\left(t^{n-1}e^{-\lambda t}\right) = 0 \Rightarrow t_{p,n}(\lambda) = \frac{n-1}{\lambda}. \qquad \text{EQ. 18.35}$$

Inserting for λ from Equation 18.24 we find

$$t_{p,n}(\Delta f) = \frac{n-1}{2 \cdot \pi \cdot \Delta f \cdot \sqrt{2n-3}}. \qquad \text{EQ. 18.36}$$

For the phase at this point of time we have

$$\begin{aligned} \phi_{p,n}(\Delta f) &= 2\pi f_0 t_{p,n}(\Delta f) \\ &= \frac{2 \cdot \pi \cdot Q \cdot \Delta f \cdot (n-1)}{2 \cdot \pi \cdot \Delta f \cdot \sqrt{2n-3}} \\ &= \frac{Q \cdot (n-1)}{\sqrt{2n-3}}. \end{aligned} \qquad \text{EQ. 18.37}$$

with $Q = \dfrac{f_0}{\Delta f}$, that is for a given choice of Q and filter order n, $\phi_{p,n}(\Delta f)$ is a constant. Now, for the detection of singularities we can make use of the fact that at $t_0 + t_{p,n}(\Delta f)$ every band will have the same phase, if a singularity at t_0 has occurred in the input signal.

Consider the expression

$$Y(t) = \left| \sum_{k=0}^{M-1} W_s(t + t_{p,n}(a_k), a_k) \right|, \qquad \text{EQ. 18.38}$$

where M is the number of frequency bands and a_k are the selected values of the scale parameter a. In words: For every frequency band take the transformation value at $t + t_{p,n}(\Delta f)$, sum them up and take the modulus of the result. At the $t = t_0$ where a singularity occurs this procedure should result in an evident peak of $Y(t)$.

This can be seen as a matched filtering procedure. Note, that the realized filter is not designed to match a particular sound, but a general onset pattern in the time-frequency plane. Strictly speaking, it matches only a single phase line

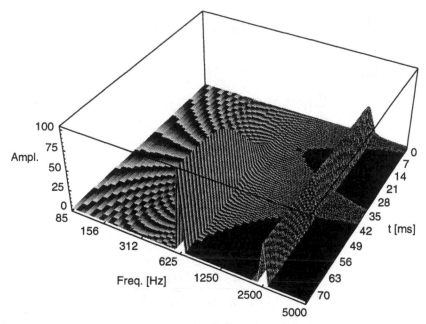

FIG. 18.6. Two cosine waves analyzed by an analytic Gaussian wavelet.

of the pattern rather than the whole one. As the given algorithm simply adds the complex wavelet moduli along this line without any additional weighing, it is optimized for $\gamma = -0.5$ as can be seen from Equation 18.33. This is a questionable choice, since it makes the filter respond optimally to singularities somewhere between the dirac and the unity step, whereas the singularities we would expect from real world signals are expected to be less dramatic. However, in the test examples we analyzed so far, this did not seem to affect the performance of the algorithm very much. It is clear though, that an optimization of the above algorithm is possible and might be necessary in certain cases.

As can be seen in Figure 18.6, the phase patterns of two sounds appearing at two different times interfere with each other. Note that each signal itself adds to the distortion of its own onset pattern. This is due to the non-homogeneity of the signals. Another point to mention is, that the distortion of the frequency axis caused by the bilinear transform also has an effect in the time domain, making Equations 18.36 and 18.37 become less and less valid for higher frequency bands. Unlike in the case of the center frequency, there does not seem to be a tractable method to find an analytic correction term, the only viable way being to compute $\phi_{p,n}(\Delta f)$ numerically by inspection of the filter responses. Again, if the sampling rate is sufficiently high this effect can be neglected.

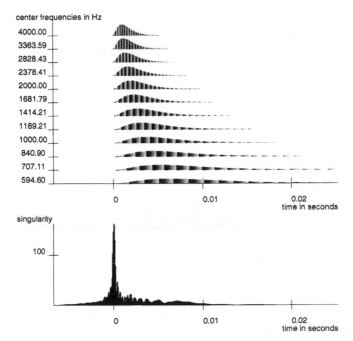

FIG. 18.7. Analytic gammatone filtering and singularity detection for the impulse function.

18.4 EXAMPLES

In this section we present two analysis examples. In Figure 18.7 we see the analytic gammatone filter output and the output of our onset detection algorithm for a unity impulse at $t = 0$. The analysis has been performed over 3 octaves descending from 4 kHz using 4 filters/octave with $\frac{\Delta f}{f_0} = 0.05$. The peak detection algorithm shows a prominent peak at $t = 0$ where the impulse occurred. The second example in Figure 18.8 shows an analysis of the first two bars of Glenn Gould's performance of the Goldberg Variations. The analysis has been performed over 3 octaves descending from 1108.73 Hz using 12 filters/octave with $\frac{\Delta f}{f_0} = 0.01$. For enhancement of the graphical display all filter output values with moduli below a certain threshold have been suppressed. Sideband crosstalk has been reduced by ignoring all filter outputs, whose instantaneous frequencies are beyond the border to the neighboring channels.

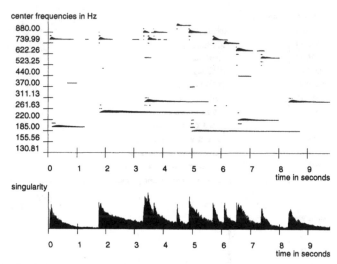

FIG. 18.8.   Analytic gammatone filtering and singularity detection, Goldberg Variations example.

Furthermore harmonics of the base frequencies have been suppressed to a great extent by a simple phase locking detection algorithm.

18.5 CONCLUSION

In this chapter links between auditory filtering and wavelet analysis were drawn. We derived an explicit formula for the time-frequency window size of the n-th order gammatone filter and showed that $n = 3$ yields the minimum group delay at the center frequency. A simple method for the detection of signal onsets was given as well as an example showing the potential of our approach for the analysis of polyphonic music.

18.6 ACKNOWLEDGMENTS

The authors gratefully acknowledge valuable hints and remarks by Richard F. Lyon.

REFERENCES

Cooke, M. (1993). *Modelling Auditory Processing and Organisation.* New York: Cambridge University Press.

Hlawatsch, F. & Boudreaux-Bartels, G. (1992). Linear and quadratic time-frequency signal representations. *IEEE SP Magazine*, pp. 21-67.

Mallat, S. & Zhong, S. (1992). Characterisation of signals from multiscale edges. *IEEE Transactions on Pattern Analysis and Machine Intelligence*, 14(7), 710-732.

Papoulis, A. (1987). *Signal Analysis.* New York: McGraw-Hill.

Patterson, R., Robinson, K., Holdsworth, J., McKeown, D., Zhang, C., & Allerhand, M. (1992). Complex sounds and auditory images. In Cazals, Y., Demany, L., & Horner, K. (Eds.), *Auditory Physiology and Perception, Advances in Biosciences*, pp. 429-443. Pergamon Press.

Rioul, O. & Vetterli, M. (1991). Wavelets and signal processing. *IEEE SP Magazine*, pp. 14-38.

Shensa, M. (1992). The discrete wavelet transform: wedding the A Trous and Mallat algorithm. *IEEE Transactions on Signal Processing*, 40(10), 2464-2482.

Simoncelli, E., Freeman, W., Adelson, E., & Heeger, D. (1992). Shiftable multiscale transforms. *IEEE Transactions on Information Theory*, 38(2), 587-607.

Slaney, M. (1993). *An Efficient Implementation of the Patterson-Holdsworth Auditory Filter Bank.* Technical report, Apple Computer Inc.

Zwicker, E. & Fastl, H. (1990). *Psychoacoustics.*, Berlin Heidelberg New York: Springer.

19

Analysis and Synthesis of Sound Textures

Nicolas Saint-Arnaud and Kris Popat
The MIT Media Laboratory

The sound of rain or of a large crowd are examples of sound textures. A restricted definition of *sound texture* is proposed for machine processing. Sound textures are treated as two-level phenomena: simple sound elements called atoms form the low level, and the distribution and arrangement of atoms form the high level. A cluster-based probability model is used to characterize the high level of sound textures. The model is then used to resynthesize textures that are perceptually similar to originals (training data). Finally, applications of the model for classification of sound textures are suggested.

19.1 DEFINITION OF A SOUND TEXTURE

In this chapter we present a method for resynthesis of sound textures, like the sound of rain, large crowds, fish tank bubbles, photocopiers and myriad others.

Defining *sound texture* is no easy task. Most people will agree that the noise of a fan is a likely "sound texture." Some other people would say that a fan is too bland, that it is only a noise. The sound of rain, or of a crowd are perhaps better textures. But few will say that one voice makes a texture.[1]

19.1.1   First Constraint in Time: Constant Long-Term Characteristics

A definition for a sound texture could be quite wide, but we chose to restrict our working definition for many perceptual and conceptual reasons. First of all, there is no consensus among people as to what a sound texture might be; more people will accept sounds that fit a more restrictive definition.

The first constraint we put on our definition of a sound texture is that it should exhibit similar characteristics over time, that is, a two-second snippet of a texture should not differ significantly from another two-second snippet. A

293

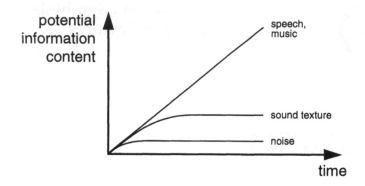

FIG. 19.1. Sound textures and noise show constant long-term characteristics.

sound texture is like wallpaper: it can have local structure and randomness, but the characteristics of the fine structure must remain constant on the large scale.

This means that the pitch should not change like that of a racing car, the rhythm should not increase or decrease, and so on. This constraint also means that sounds in which the attack plays a great part (like many timbres) cannot be sound textures. A sound texture is characterized by its sustain.

Fig. 19.1 shows an interesting way of segregating sound textures from other sounds, by showing how the "potential information content" increases with time. "Information" is taken here in the cognitive sense rather then the information theory sense. Speech or music can provide new information at any time, and their "potential information content" is shown here as a continuously increasing function of time. Textures, on the other hand, have constant long term characteristics, which translates into a flattening of the potential information increase. Noise (in the auditory cognitive sense) has somewhat less information than textures.

Sounds that carry a lot of meaning are usually perceived as a message. The semantics take the foremost position in the cognition, downplaying the characteristics of the sound proper. We choose to work with sounds which are not primarily perceived as a message, that is, nonsemantic sounds, but we understand that there is no clear line between semantic and non-semantic. Note that this first time constraint about the required uniformity of high level characteristics over long times precludes any lengthy message.

19.1.2 Two-Level Representation

Sounds can be broken down to many levels, from a very fine (local in time) to a broad view, passing through many groupings suggested by physical, physiological and semantic properties of sound. We choose, however, to work with only two levels: a low level of simple atomic elements distributed in time and a high level describing the distribution in time of the atomic elements.

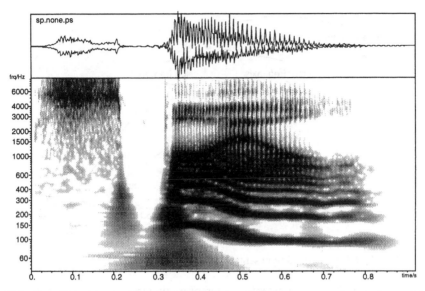

FIG. 19.2. Example of a time-frequency representation: a constant-Q transform of the word "spoil." (Courtesy of Dan Ellis.).

For many sound textures - applause, rain, fish-tank bubbles - the sound atom concept has physical grounds. Many more textures can also be usefully modeled as being made up of atoms. Without assuming that all sounds are built from atoms, we use the two-level representation as a model for the class of sound textures that we work on.

The boundary between low and high level is not universal or fixed, and we will in fact move it, by sometimes using very primitive atomic elements, sometimes using more complex atoms. Note that using simpler atoms leaves the high level to deal with more information and more complexity. On the other hand, one should be careful not to make overly narrow assumptions — losing generality — when choosing more complex atomic elements.

Such a two-level representation has some physical grounding, as explored in "Auditory Perception of Breaking and Bouncing Events" (Warren and Verbrugge, 1988). In this paper, Warren and Verbrugge present a "structural" level characterized by the properties of the objects being hit, and a "transformational" level, characterized by the pattern of successive hits in breaking and bouncing events.

19.1.3 Low Level: Sound Atoms

The signal captured by a microphone is a time waveform, which can be digitally represented by Pulse Code Modulation (PCM) (Reddy, 1976). In the human ear, the cochlea performs a time-frequency transform. Fig. 19.2 shows the time waveform for an occurrence of the word "spoil", and an example of time-

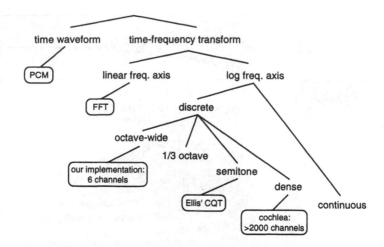

FIG. 19.3. A variety of sound representations.

frequency transform (a constant-Q transform) underneath. A time-frequency representation is often called spectrogram.

Fig. 19.3 shows a variety of sound representation domains, including the time waveform (PCM) and a few possible time-frequency diagrams.

Sound atoms usually form patterns in one of the representation domains. For example, the time-frequency representation of the phonemes of the word "spoil" on Fig. 19.2 could be used as atoms. The atom patterns can be complex, but we can also choose simpler atoms, such as groupings of energy continuous in time and frequency. Finally, the instantaneous energy in each frequency channel of a time-frequency transform form the simplest atomic features possible.

Instantaneous Energy. In our implementation for resynthesis, we choose to use the simplest atomic elements: instantaneous energy in one frequency channel. In our implementation, a straightforward filter bank splits the incoming sound in six octave-wide frequency bands. The energy levels in each band are our atomic "features." The small number of frequency bands is convenient for computation, as we will see in the implementation.

In the limit, with one frequency band (no filtering), atoms can be chosen as the instantaneous PCM value in the broadband signal. This simplistic approach shifts the computational burden to the high level. The "two sines" resynthesis example in section 19.3.3 uses the PCM values as atoms.

Using narrower filters, perhaps semi-tone-spaced like Ellis' constant-Q filter-bank (Ellis, 1992), yields a transformation closer to human hearing. However, the amount of data is greatly increased, and the increased number of channels makes processing more complex than in our current system. A new system tailored for multiple channels is worth exploring in future work.

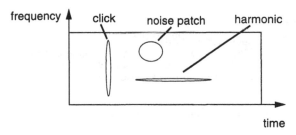

FIG. 19.4. Clicks, harmonic components and noise patches on a spectrogram.

Harmonic Components, Clicks and Noise Patches. On a spectrogram, the energy is not uniformly distributed, but tends to cluster. Grouping together energy that is adjacent in time and frequency, and parameterizing these groupings allows a great reduction of the amount of data. Fig. 19.4 shows three possible groupings of energy adjacent in time and frequency.

Musical instruments and speech show tracks on their narrow-band spectrograms, that is, lines of one frequency with a continuity in time. Smith and Serra (1987) describe a method to extract tracks from a short-time Fourier transform (STFT) spectrogram. Ellis obtains tracks from his constant-Q spectrogram (Ellis, 1992).

Tracks describe harmonic components, but are not well suited for clicks (broadband, short time) or noise patches (energy spread over both time and frequency). For those, a richer representation is required that allows atoms to span both time and frequency intervals. Matching Pursuit (Mallat and Zhang, 1992) is a method that can possibly extract such diverse atoms from sounds.

19.1.4 High Level: Distribution of Atoms

The high level of our two-level sound representation is concerned with the distribution of the sound atoms extracted at the low level. We identify periodic and stochastic (random) distributions of atoms, as well as co-occurrence and sequences of atoms. These different ways of distributing atoms are not exclusive of each other; they can be mixed and even combined in a hierarchy.

A sound in which similar atoms occur at regular interval in time is said to have a periodic distribution. Textures such as engine sounds have a periodic distribution. In a stochastic distribution, atoms occur at random times but obey some arrival rate. Rain and applause are examples of textures with a stochastic distribution of atoms. Different atoms that occur at the same time are said to co-occur. The impact of objects makes a sound where atoms co-occur in different frequency bands. Atoms also often occur in predictable sequences.

As an example, our photocopier makes a sucking sound in which many frequency components have high energy (co-occurrence). The sucking sound is followed (sequence) by a feeder sound. The suck-feed sequence is repeated

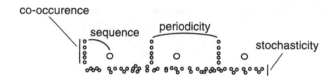

FIG. 19.5. Example of distributions of atoms: the Copier.

sequentially (periodicity). At all times there is a low rumble (stochasticity). Fig. 19.5 is a stylized representation of those four kinds of distributions.

Occurrences of atoms can be grouped into a hierarchy, for example, the sucking sound (a co-occurrence) is periodic. The high level model should address all four kinds of distributions, as well as hierarchic distributions of distributions.

Unified Description of Distribution. The method we use for characterizing the distribution of atoms (the cluster-based probability model) does not assume that the texture takes a particular distribution. Instead, it tries to characterize the distribution of atoms by keeping statistics on the most likely transitions.

19.1.5 Second Time Constraint: Attention Span

The sound of cars passing in the street illustrates an interesting problem: if there is a lot of traffic, people will say it is a texture, whereas if cars are sparse, the sound of each one is perceived as a separate event. We call *attention span* the maximum time between events before they become distinct. A few seconds is a reasonable attention span, but once again, there is no sharp boundary.

We therefore put a second time constraint on sound textures: high-level characteristics must be exposed or exemplified (in the case of stochastic distributions) within the attention span of a few seconds.

This constraint also has a good computational effect: It makes it easier to collect enough data to characterize the texture. By contrast, if a sound has a cycle of one minute, several minutes of that sound are required to collect a significant training set. This would translate into a lot of machine storage, and a lot of computation.

19.1.6 Summary of Working Definition of Sound Texture

(1) Sound textures are formed of basic sound elements, or atoms;

(2) atoms occur according to a higher-level pattern, which can be periodic, random, or both;

(3) the high-level characteristics must remain the same over long time periods (which implies that there can be no complex message);

(4) the high-level pattern must be completely exposed within a few seconds ("attention span");

(5) high-level randomness is also acceptable, as long as there are enough occurrences within the attention span to make a good example of the random properties.

19.2 A METHOD FOR HIGH-LEVEL CHARACTERIZATION: THE CLUSTER-BASED PROBABILITY MODEL

To characterize the high-level transitions of sound atoms, we use the cluster-based probability model (Popat and Picard, 1993). This model summarizes a high dimensionality probability mass function (PMF) by describing a set of clusters that approximate the PMF. Popat and Picard have used the cluster-based probability model for visual textures and image processing.

19.2.1 Overview

The cluster-based probability model encodes the most likely transitions of ordered features. Features (in our case sound atoms) are put in vectors, and the order within the vector reflects a pre-established order of features in time and frequency. The features used to encode the transitions are taken in the neighborhood of the current feature; therefore we call the set of the relative positions of the conditioning features a neighborhood. The vectors formed from a training sound are clustered to summarize the most likely transitions of atoms.

The input to the analyzer are a series of vectors in N-dimensional space representing the features of the training data. The vectors are clustered using a K-means algorithm (Therrien, 1989), slightly modified to iteratively split its clusters. The centroid of each cluster, its variances and relative weight then form a lower-dimensionality estimate of the statistics of the training vectors. The next section describes the use of a cluster-based probability model to characterize transitions of features, and its application to characterization of sound textures for resynthesis.

19.2.2 Application to sound texture characterization

The model attempts to characterize a sound texture, which we call the *training signal*. First the sound must be put in digital form, by sampling the analog signal. The simplest features for an audio signal are the digital values output by the analog-digital process. To study the transitions of features, we build vectors of features ordered in time but not necessarily adjacent or even equally spaced in

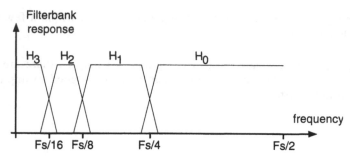

FIG. 19.6. Frequency responses for the 3- level tree-structured QMF filter bank.

time. Think of a ruler with holes punched in it; when you put the ruler over the sampled signal, you see values in each hole: This string of values is a feature vector. (Later we work with vectors spanning both time and frequency, and the 1-D ruler will become a 2-D neighborhood mask.) Different feature vectors can be obtained by sliding the ruler. If you slide the ruler over the length of the training signal, you obtain a set of training vectors. We use this set of training vectors to demonstrate the cluster-based probability model.

Each vector in the training set is an observed transition of features in time. Given that the training set is large enough to contain many observations of each likely transition, we can build a model that clusters the likely transitions and assigns a probability to each.

Each transition (training) vector is a point in a d-dimensional vector space. We assume that the signal is characterized by a probability mass function (PMF) in this vector space. Then the training vectors are samples of this PMF. Likely feature transitions correspond to dense areas of the PMF; we summarize the likely transitions by identifying clusters of points in vector space. For each cluster, we compute the center and variance of the points, and we assign a probability for the cluster based on the number of points, that is, we estimate the PMF by "histogramming" the samples of the process.

Now suppose that we have a piece of a signal, and that we want to add one likely feature at the end. To synthesize this new feature value we need to estimate its PMF conditioned on the feature values preceding it. By placing the vector-building ruler so that the last hole faces the desired new feature position, a vector identifies the desired region of space for d-1 dimensions. We compute the effect of each cluster along the unknown axis to get an estimate of the distribution of the value (one-dimensional PMF) at that last position. The process can be repeated to resynthesize long portions of signal.

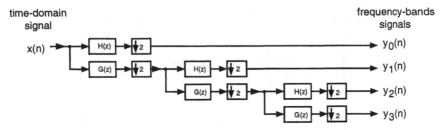

FIG. 19.7. A 3-level tree structured QMF filter bank.

19.3 RESYNTHESIS OF SOUND TEXTURES

19.3.1 Low-Level: Atom Extraction

We choose to do a frequency decomposition of the incoming sound using a binary tree structured QMF (Quadrature Mirror Filter) filter bank (Vaidyanathan, 1993). This filter bank shows a great simplicity in design. Each band (except the lowest) is one octave wide, so that the bandwidth halves at every stage, which satisfies the constant-Q criterion (Fig. 19.6). An example of a simple tree structured QMF filter bank structure is shown in Fig. 19.7.

19.3.2 High Level: Transitions of Atoms

Dimensionality Problem. Each point in the feature vector corresponds to a dimension in PMF space. In order to get a reasonable estimate of the PMF, the space must be well covered by training vectors. The more dimensions this space has, the more difficult it becomes to fill it with training vectors. We kept the number of dimensions to a maximum of 14.

Because each point in the conditioning neighborhood adds one dimension in our space (many if we use a frequency transform), we see clearly that the number of points in the neighborhood has to be kept as small as possible.

Because the transitions of atoms are not the same in all frequency channels, a different model is used for each channel in the current implementation. This multiplies the overall computational complexity, and it is another reason to keep the size of each model to a minimum.

Neighborhood Masks. To build the neighborhoods, we used only values from the past (causality), and we estimate roughly that a sample can be correlated to previous samples up to a second before.

For a given data point, the most basic neighborhood is the previous data in the same frequency band. A feature depends a lot on the values that

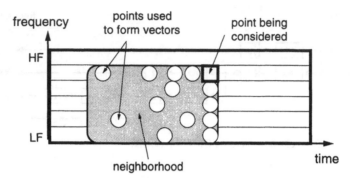

FIG. 19.8. Neighborhood on a 6-band time-frequency diagram.

immediately precede it, so some points in the close past should be part of the mask. However, to capture long-time transitions, the mask should also include some points further in the past. The general rule we use is that the points should be spaced further apart as the delay in the past increases, to keep the number of points small. In Fig. 19.8, those are the circles on the top horizontal line.

Another important part of the neighborhood is made up of the points that occur simultaneously in other frequency bands. These form the vertical line of Fig. 19.8. If the number of bands is low (as in our case), we can use a point from each band.

During resynthesis, the neighborhood should use only points already generated. Since the signal is resynthesized channel by channel starting at the low frequencies, the neighborhoods we used had atoms only in the same frequency band or lower frequency bands, as in Fig. 19.8.

19.3.3 Results

Two sines. As a reality check on the model, we tried a signal composed of two sine waves of non-integer frequency ratios as a simple test of the analysis-synthesis cycle. This signal, as all others we used, is sampled at 22.05kHz. We did not use any filtering; the features were chosen as the instantaneous sampled value of the signal. A short run of the original signal and the resynthesis are shown on Fig. 19.9.

The model has captured most of the sound; one can see that the waveform is reproduced, with some glitches. The resynthesized signal sounds like the training signal with a aliasing-sounding high-frequency hiss superposed.

The glitches could be due to an insufficient number of clusters to code all possible transitions: the signal would err for a few samples before the neighborhood would again correspond to an existing cluster.

The signal exhibits a marked start-up stage before becoming stable, as can be seen on Fig. 19.10. This is due to the absence of points to fill the neighborhood at the beginning: the model starts in a random state, and

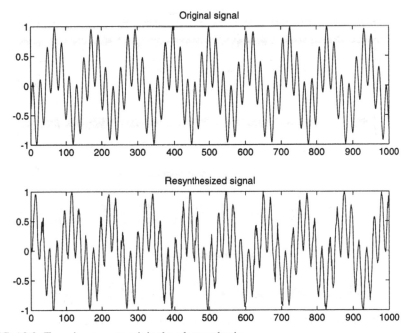

FIG. 19.9. Two sine waves: original and resynthesis.

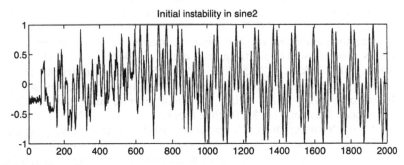

FIG. 19.10. Two sine waves: start-up stage.

eventually reaches known transitions. This problem becomes much more severe with more complex sounds, as we will see in the photocopier example.

Photocopier. Our next resynthesis attempt was done on a photocopier sound that consists of a dominant fan noise with repetitive feeder noises. The fan noise has its energy spread unequally among frequencies, like a white noise passed through a filter with resonances. The feeder sound is similar to a low frequency burst followed by a broadband click occurring at roughly equal time delays (see Fig. 19.5. Details of both the waveform and the spectrograms of the copier

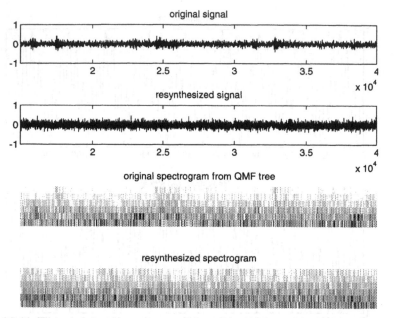

FIG. 19.11. Time signal and spectrogram of the original and resynthesized copier signal.

example can be seen in Fig. 19.11; the clicks occur around 1.7, 2.5 and 3.3x10^4 in the time scale. (Note that the time scale is subsampled by 2, so there are 11k spectrogram slices per second).

Setting the parameters was the primary concern. For our example, the filter bank was chosen to have six frequency bands: 0-345Hz, 345-691Hz, 691Hz-1.38kHz, 1.38-2.76kHz, 2.76-5.5kHz, 5.5-11kHz (these are dependent on the sampling frequency of 22.05kHz). For each frequency band, we took about 8000 vectors (every third point) of N=14 points, and M=128 clusters. Reconstruction was done for 51200 time slices.

The neighborhoods for each frequency band chosen look very much like that in Fig. 19.8. For the lower frequency bands, the points that could not be allocated to low frequencies were used to increase the number of points in the past of the same frequency band, so that the total number of points in all neighborhood masks was constant at 13 (current point + 13 = 14 = N).

The resynthesis has kept most of the flavor (resonances) of the noise, but there are many more clicks, and they come at very irregular intervals (see Fig. 19.11). At the beginning of the resynthesis, there is no neighborhood present, so the model generates a transitory signal. We suspect that in this case, it has generated many feeder-clicks early on, and all these clicks are repeated periodically, so that there are many clicks in a time interval that should contain only one feeder sound.

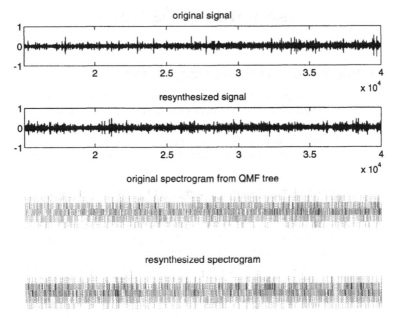

FIG. 19.12. Applause example.

Applause. A more convincing resynthesis was made from the sound of a crowd applauding. The atom extraction, model size and neighborhood masks are the same as those for the copier example. The resynthesis is a bit rough, but recognizable as applause. The spectrogram is shown on Fig. 19.12. A sound sample is available from http://www.media.mit.edu/~nsa/IJCAI-95.html.

Analysis of Results. The signals that were successfully resynthesized were all very stochastic in nature, with the exception of the two sines, which are deterministic. Attempts to resynthesize signals with periodicities in the order of hundreds of milliseconds, such as the sound of an helicopter or of a motorcycle, were much less successful. The resynthesis for these had a similar frequency content, and even a similar "flavor" to them as the originals, but the long-time periodicity was not properly reproduced.

All the resyntheses used a six channel filterbank. This is certainly too crude an approximation for the cochlea. Using the instantaneous amplitude in each filter channel is also a very crude feature selection. Nonetheless, the model has been able to resynthesize sound textures with some success. This leads us to believe that the model is useful and that the resynthesis could be improved by refining the atom selection and taking care of the start-up conditions.

19.4 FUTURE DIRECTIONS

19.4.1 Improvements to the Resynthesis

Increasing the dimensionality of the models used may seem like a way to model more transitions and interrelations between atoms, and thus improve analysis. However, it could be argued that if the features - atoms - being fed into the model are not good at representing the training data, then the model will not be able to summarize the transitions, however large the dimensionality of the model.

Better Atoms. In the "applause" example, the "natural" atoms are individual hand claps; is the analysis method good for characterizing this? If we were to choose individual hand claps as atoms, we would need both a "clap extractor" and a "clap resynthesizer" - which could be simply a single hand clap sample player. The resynthesis would most probably greatly improve, but the analyzer-resynthesizer combination would perform poorly on textures other than applause.

The main direction of improvement lies in choosing more meaningful atoms, and changing the atom extraction method accordingly. Somewhere between the simplistic "energy in a frequency band" atom and the specific "single hand clap" atom lies a set of physically meaningful yet general atoms. Do "harmonics, clicks and noise patches" (in section 19.1.3) form such a set?

Start-Up Condition. The initial condition is often problematic for resynthesizing sound textures: in the start-up stage, the neighborhood is mostly empty, which brings the model to very sparsely populated areas of the PMF space; the model might never get to the dense areas of the space. For those textures where the initial condition is important, it is possible to provide a seed -- a snippet of the original texture to build the first neighborhoods.

Experiments have shown that textures with strong periodic components - helicopters, motorcycles - do not resynthesize well. These textures are also very likely to be hurt by an improper start-up. Providing a sine wave of the fundamental frequency of the training texture as an initial condition might be a way to help the model converge in the right region of space for those textures.

Computation Speed. Another area requiring improvement is the computation speed. Current times on a DEC Alpha are roughly one hour of computation time for each second of resynthesized sound. Most of the machine time is spent computing the effect on the PMF of clusters that are so far from the current quadrant in the vector space that their contributions are negligible. Performance could be largely improved if those computations were avoided from the start. Preliminary results show that eliminating negligible clusters can speed up execution by a factor of 5 (Popat and Picard, 1997).

19.4.2 Classification of Sound Textures

There are many other interesting applications of the sound texture model that revolve around classification and identification. All involve comparing an unknown texture with one or more known "template" textures. Here are a few simple classification problems:

(1) compare two textures

(2) classify a texture as belong to one of several classes, possibly with confidence level for the result.

(3) classify a texture as belong to one of several perceptual classes, each with multiple templates (e.g. watery sounds have templates for rain and bubbles).

All the preceding problems require a distance measure between textures. With the cluster-based probability model, we distinguish two main ways to compare two textures, resulting in two different distance measures:

(1) compare the transitions of atoms in the unknown texture with the model for the known texture (model fit, e.g., Bayesian classification), and

(2) compare the model extracted from the unknown texture with the model for the known texture (model comparison).

Both classification schemes assume that sound atom extraction is performed in the same way on all textures (unknown and templates), and that transition vectors are built with the same neighborhoods. Bayesian classification is straightforward and directly provides a probability that the unknown texture fits the model, but provides no insights about how the unknown and the template differ. Comparing models is much more complex because there is no obvious practical distance measure between models. However, comparing models could lead to discovery of how global parameters of texture - like periodicity - are expressed in the models.

19.5 CONCLUSION

We have presented a restricted definition of a sound texture for machine processing, which requires long-term stationarity of the characteristics of the sound. We model the textures as being composed of two levels: simple sound atoms and the distribution of the atoms. A cluster-based probability model is used to characterize the high level of sound textures. The model is then used to resynthesize textures that are in some cases similar to originals. The main source of improvement to the resynthesis is thought to be in a better selection of the type of atoms. Finally, we propose to use the model for classification of sound textures.

REFERENCES

Ellis, D. (1992). *A Perceptual Representation of Audio.* Cambridge, Massachusetts: Master's thesis, Department of Electrical Engineering and Computer Science, Massachusetts Institute of Technology.

Mallat, S. & Zhang, Z. (1992). *Matching Pursuit with Time-Frequency Dictionaries.* New York: Courant Institute of Mathematical Sciences. (Technical Report No. 619)

Popat, K. & Picard, R. (1993). A novel cluster-based probability model for texture synthesis, classification, and compression. Cambridge, Massachusetts: *Proceedings of SPIE Visual Communications '93.*

Popat, K. & Picard, R. (1997). *Cluster-based probability model and its application to image and texture processing.* IEEE Transactions on Image Processing, February 1997 (to appear).

Reddy, D. (1976). Speech recognition by machine: A review. *IEEE Proceedings*, 64, 502-531.

Smith, J. M. & Serra, X. (1987). Parshl: an analysis/resynthesis program for non-harmonic sounds based on a sinusoidal representation. In *Proceedings of the 1987 ICMC*, p. 290ff.

Therrien, C. (1989). *Decision, Estimation and Classification.* New York: Wiley.

Vaidyanathan, P. (1993). *Multirate Systems and Filter Banks.* Englewood Cliffs, New Jersey: Prentice-Hall.

Warren, W. & Verbrugge, R. (1988). Auditory perception of breaking and bouncing events. In Whitman Richards (Ed.), *Natural Computing.* Cambridge, Massachusetts: MIT Press.

NOTE

[1] Except maybe high-rate Chinese speech for someone who does not speak Chinese.

20

Predicting the Grouping of Rhythmic Sequences using Local Estimators of Information Content

Steven M. Boker
University of Notre Dame

Michael Kubovy
University of Virginia

The hypothesis is proposed that auditory events are perceived to be partitioned according to boundaries constructed at times of maximum surprise. One way of quantifying surprise is via information theoretic predictions. The results of two experiments are presented that test the plausibility of this hypothesis using simple repeating auditory rhythmic sequences. Local estimators of information content within an auditory sequence are used to construct predictors of perceived segmentation. These predictors are fit to results of the experiment by using a structural equation model and are compared with Garner's Run-Gap model (Garner, 1974). The information theoretic model is found to be a significantly better predictor of the experimental results than the Run-Gap model.

20.1 INTRODUCTION

The segmentation and ordering of a continuous sensory stream into a series of recognizable events presents one of the fundamental problems in perception (Lashley, 1952). The auditory system must partition the incoming stream in a meaningful way; one that preserves relationships within the stream, but also

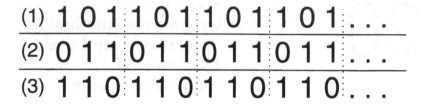

FIG. 20.1. A repeating sequence of length 3 has 3 potential starting points.

breaks the stream into a sequence and thus allows the recognition of words, phrases or sentences.

In music, these sequential units are formalized and regularized in such a way that much of the possible ambiguity in segmentation is removed. This process of disambiguation is achieved in a variety of ways, primarily by stress or accent (Handel, 1989). Why is there a need to disambiguate the structure of such regular musical patterns? The rhythmic structure of spoken language carries content that helps the listener organize the sounds of speech into grammatical sense during the process of comprehending the meaning of the sentences (Thomas, Hill, Carrol, & Garcia, 1970). Why is there a need for prosodic elements in speech?

By studying the nature of the ambiguity in rhythmic sequences we can understand and predict the organization which will be perceived to be inherent in the timing of the sequence. This will lead to a more precise understanding of the interrelationships between stress and timing which create the unambiguous perception of segmentation of auditory streams.

A simple repeating rhythmic pattern which contains no stressed elements may be perceived as having a variety of starting points. Figure 20.1 shows a repeating sequence that could be perceived as having one of three potential starting points. Some starting points have a higher probability of being perceived than others, but each of these probabilities is greater than zero.

Garner and his colleagues (Royer & Garner, 1966; Garner & Gottwald, 1968; Garner, 1974) studied these types of rhythmic patterns and devised heuristics, which they named the *run principle* and *gap principle*, by which predictions could be made regarding the organization that would be perceived by individual subjects. The work presented here replaces Garner's heuristics with a more formal information theoretic (Shannon & Weaver, 1949) estimation of the probability of perceiving any starting point as a segmentation boundary. This relationship between local information content in the perceptual stream and the perception of temporal segmentation is likely to generalize to the other sensory modalities.

20.2 METHODS

20.2.1 Experiment 1

Subjects. Eleven subjects participated in Experiment 1, 8 males and 3 females. Age of the subjects ranged from 18 to 41. Six subjects reported having received more than 4 years of training in playing a musical instrument, whereas the remaining five subjects reported no formal training in playing a musical instrument. All but two subjects reported being right-handed.

Experimental Procedure. Each subject was asked to complete the experimental procedure once on each of five occasions, where the occasions were separated by as little as 24 hours and by as much as three weeks. Each session required approximately 45 minutes to complete.

In each trial, subjects were presented with one repeating rhythmic auditory pattern and were asked to respond by striking a key on a synthesizer keyboard synchronous with the perceived starting point of the pattern. Subjects were asked to continue to strike the key at the beginning of each repetition until confident that they had perceived the starting point. Once subjects were confident of their response, they were asked to press a mouse button ending the trial.

The PsyLog software (Boker & McArdle, 1992), running on a NeXT workstation, was modified to present the stimuli and gather the responses. Subjects were presented with a short set of practice trials and then a set of 115 experimental trials. Subjects were informed that they could rest at any time that they became fatigued and that the software would wait for them.

A single rhythmic stimulus was composed of a fixed number of *beats*, equal intervals of time which could either be empty or be filled with a percussive sound at the beginning of the interval. Thus, a measure could be represented by a binary number where each binary digit represents a beat: a zero representing an empty interval and a one representing an interval with a percussive sound at its beginning. The set of stimuli for Experiment 1 consisted of all of the unique rhythmic patterns of length 8 or less, 115 patterns in all.

In each trial, the stimulus initially began with a short beat length of 10 ms, which quickly slowed to a steady beat length of 250 ms. The beat on which the stimulus was initiated was chosen at random for each presentation of the stimulus. The combination of these two methods minimized the subjects' ability to associate the beginning of the presentation of the first beat of the stimulus with the perceived beginning of a measure within the repeating pattern.

The percussive sound used in this experiment was a synthesized musical cowbell produced by a Roland MT-32 MIDI wavetable synthesis module and delivered to the subjects binaurally via Sennheiser HD-414-SL headphones. The user responded by striking a key on a Kawai K5 Digital Synthesizer keyboard.   Several variables were measured for each keypress: the time of response in milliseconds relative to the beginning of the presentation of the stimulus, the time of response in milliseconds relative to the beginning of the pattern as represented internally by the computer software, and the velocity of the response as an integer between 1 and 127.

20.2.2 Experiment 2

Subjects. Twenty eight subjects participated in Experiment 2, 17 males and 11 females. Age of the subjects ranged from 18 to 21. Thirteen subjects reported more than 4 years of training in playing a musical instrument. Sixteen subjects reported being right-handed.

Experimental Procedure. The experimental procedure was identical to that of Experiment 1 with the following two exceptions. The subjects in Experiment 2 were only tested on one occasion. The set of rhythmic stimuli in Experiment 2 consisted of a random sample of 115 stimuli drawn half from the unique patterns of length 8 or less and half from the unique patterns of length 12.

20.3 MODELS

20.3.1 Run-Gap Predictions

A latent variable structural equation model was employed to test the goodness of fit of Garner's "run-gap" heuristic predictions to the data gathered from the two experiments. Figure 20.2 shows a path model of Garner's run-gap heuristics. The predictor variables are *Run*, the run principle, and *Gap*, the gap principle. The latent variable is S, the perceived structure of the rhythmic pattern. The measured outcome variables are RB, the response within the beat; A, the accuracy of the response; and V, the velocity of the response.

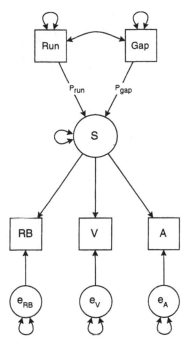

FIG. 20.2. Path diagram showing a structural model of Garner's basic run-gap theory.

The Run variable is constructed as follows: if the current stimulus beat value is 1 and the previous stimulus beat value is 0, then Run = 1 + the number of stimulus beats following the current stimulus beat before another stimulus beat with a value of 0 is encountered. The Gap variable is constructed similarly: if the current stimulus beat value is 1 and the previous stimulus beat value is 0, then Gap = 1 + the number of stimulus beats preceding the current stimulus beat before a stimulus beat with a value of 1 is encountered.

The outcome variables were coded as follows. If a keypress occurred within ±.5 of the onset of the current stimulus beat, then *RB* was coded as 1, otherwise as 0. If *RB* was 1, the MIDI velocity of the keypress (range 1-127) was coded as *V*. If RB was 1, *A* was coded as

$$A = 1 - \frac{2(t_k - t_b)}{b}$$

where t_k is the elapsed time to the keypress, t_b is the elapsed time to the onset of the stimulus beat, and b is the duration of the beat. This means that the accuracy of the keypress was 1 if the keypress occurred simultaneously with the onset of the stimulus beat, and the accuracy was 0 if the keypress occurred halfway between two stimulus beats.

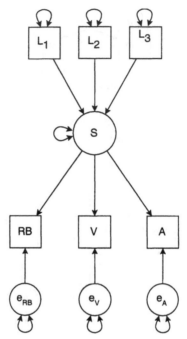

FIG. 20.3. Path diagram of entropy prediction model where the predictors are: L1 , entropy of features of length 1; L2 , entropy of features of length 2; and L3 , entropy of features of length 3.

RM was coded as 0 if no keypress occurred within 0.5 measure of the current stimulus beat. Thus *RM* began the trial as a string of 0's and became a string of 1's when the subject began to respond to the stimulus.

20.3.2 Local Information Predictions

A similar latent variable model was constructed to test the fit of the local information content predictions to the data from Experiments 1 and 2 (see Figure 20.3). The outcome variables were coded in exactly the same manner as for the Run-Gap Model above. The predictor variables were calculated in terms of redundancy for features of a particular size.

Redundancy can be stated in terms of a ratio of entropies (Barlow, 1961), and was explored by Redlich (1993) as an active mechanism in visual perception. In the case of segmentation, the quantity which needs to be calculated is a measure of local information content rather than redundancy. The measure of local information content which will be used here is simply 1 - *R*, where *R* is redundancy.

Consider a repeating sequence of beats. If r is the number of contiguous repetitions with which a feature x of size s has previously occurred, then one possible measure of local surprise upon the reoccurrence of x is

$$L_x = \frac{H_x}{H_{\sim x}}$$

$$= \frac{-r/(r+1)\log_2(r/(r+1))}{-1/(r+1)\log_2(1/(r+1))}$$

$$= \frac{r\log_2(r/(r+1))}{\log_2(1/(r+1))},$$

whereas if x does not re-occur then

$$L_x = \frac{H_{\sim x}}{H_x}$$

$$= \frac{\log_2(1/(r+1))}{r\log_2(r/(r+1))}.$$

For each beat in the rhythmic sequence, three local measures of information content were calculated L_1, L_2, and L_3. These three predictor variables have a value for each beat of each rhythmic sequence.

20.4 RESULTS

The two models were fit to the data from Experiment 1 and Experiment 2 using the structural equation modeling procedure in SAS (PROC CALIS). The results of fitting the Run-Gap Model and the Local Information Model are presented in Table 20.1 and Table 20.2. The estimated parameters and χ^2 statistics are presented side by side for the two models.

Table 20.1. Comparison of prediction model parameters and χ^2 for models fit to the data from Experiment 1.

	Run–Gap	Entropy	
Run→S	0.043	0.295	$L_1 \rightarrow$S
Gap→S	0.131	-0.056	$L_2 \rightarrow$S
		-0.132	$L_3 \rightarrow$S
S→RB	=1	=1	S→RB
S→V	0.539	0.543	S→V
S→A	0.618	0.614	S→A
eRB→RB	-0.025	0.021	eRB→RB
eV→V	0.032	0.032	eV→V
eA→A	0.107	0.108	eA→A
Run↔Gap	0.603	0.009	$L_1 \leftrightarrow L_2$
		0.001	$L_1 \leftrightarrow L_3$
		0.000	$L_2 \leftrightarrow L_3$
Var(Run)	0.990	0.091	Var(L_1)
Var(Gap)	0.965	0.029	Var(L_2)
		0.015	Var(L_3)
Var(S)	0.105	0.124	Var(S)
χ^2	2623	79	χ^2
DF	4	6	DF
N	174781	174145	N

In Table 20,1 the Run-Gap model has a χ^2 fit statistic of 2623 with 4 degrees of freedom. Although this might seem large, recall that the effective sample size is 174,781 separate stimulus response pairs which contribute to a null model χ^2 of 923,649 with 10 degrees of freedom. The Run-Gap model certainly fits much better than the null model. However, notice that the Local Information (Entropy) Model has a χ^2 of only 79 with 6 degrees of freedom. Although these two models are not nested and so cannot be compared precisely in terms of χ^2 goodness of fit, the difference in χ^2 is so great that it overwhelms the possible loss of accuracy due to the non-nested nature of the comparison.

Table 20.2 shows the results of the analysis on Experiment 2. The Run-Gap model has a χ^2 fit statistic of 388 with 4 degrees of freedom compared to a null model χ^2 of 541,596 with 10 degrees of freedom. The Local Information

Table 20.2. Comparison of prediction model parameters and χ^2 for models fit to the data from Experiment 2.

	Run–Gap	Entropy	
Run→S	0.021	0.161	$L_1 \to$S
Gap→S	0.082	0.031	$L_2 \to$S
		-0.038	$L_3 \to$S
S→RB	=1	=1	S→RB
S→V	0.533	0.533	S→V
S→A	0.595	0.595	S→A
eRB→RB	0.039	0.040	eRB→RB
eV→V	0.031	0.031	eV→V
eA→A	0.092	0.092	eA→A
Run↔Gap	0.779	0.021	$L_1 \leftrightarrow L_2$
		0.004	$L_1 \leftrightarrow L_3$
		0.004	$L_2 \leftrightarrow L_3$
Var(Run)	1.302	0.127	$\mathrm{Var}(L_1)$
Var(Gap)	1.284	0.037	$\mathrm{Var}(L_2)$
		0.014	$\mathrm{Var}(L_3)$
Var(S)	0.085	0.093	Var(S)
χ^2	388	59	χ^2
DF	4	6	DF
N	104512	104512	N

(Entropy) Model has a χ^2 of 59 with 6 degrees of freedom. Again these two models aren't nested, but the difference in χ^2 between the two models is large enough that the Local Information Model is preferred.

20.5 DISCUSSION

The results of these analyses suggest that predicting segmentation boundaries in auditory streams by using local estimators of information content may result in calculated segmentations of temporal structure that mimic those perceived by human listeners. An algorithm estimating information content using ratios of entropies could be implemented in a computationally efficient manner since the algorithm only requires two $\log_2$ operations, one divide and a few lookups. The algorithm has the potential for parallel implementation in that the components for each feature size are calculated independently and then linearly combined.

The Local Information Model predicts that the probability of a temporal segmentation occurring at any point is directly related to the amount of information in the auditory stream at that point in time. This prediction is difficult to verify without a specific measurement model for estimating the local information in a stream on a moment by moment basis. Finding a reliable estimator for local information in complex stimuli remains an open problem, but a problem which if solved seems likely to hold rewards for research in auditory scene analysis.

20.6 ACKNOWLEDGMENTS

We would like to thank Jay Friedenberg whose help in performing this experiment was invaluable. We would also like to thank several anonymous reviewers whose comments and criticisms have substantially improved this chapter. This work was supported by NIMH grant number 5-R01 MH47317.

REFERENCES

Barlow, H. B. (1961).   Possible principles underlying the transformation of sensory messages. In Rosenblith, W. A. (Ed.), *Sensory Communication*, pp. 217-234. Cambridge, MA: MIT Press.

Boker, S. M. & McArdle, J. J. (1992).   PsyLog: Software for psychometric measurement.   Unpublished software, Department of Psychology, University of Virginia.

Garner, W. R. & Gottwald, R. L. (1968).   The perception and learning of temporal patterns. *Quarterly Journal of Experimental Psychology, 20*, 97-109.

Garner, W. R. (1974). *The Processing of Information and Structure*. Hillsdale, NJ: Lawrence Erlbaum Associates.

Handel, S. (1989).   *Listening: An Introduction to the Perception of Auditory Events*. Cambridge, MA: MIT Press.

Lashley, K. S. (1952).   The problem of serial order in behavior.   In Jeffress, L. A.(Ed.), *Cerebral Mechanisms in Behavior,* pp. 112-136.   New York: Wiley.

Redlich, N. A. (1993).   Redundancy reduction as a strategy for unsupervised learning. *Neural Computation*, 5, 289-304.

Royer, F. L. & Garner, W. R. (1966).   Response uncertainty and perceptual difficulty of auditory temporal patterns. *Perception & Psychophysics*, 1, 41-47.

Shannon, C. E. & Weaver, W. (1949).   *The Mathematical Theory of Communication.*   Urbana: The University of Illinois Press.

Thomas, I. B., Hill, P. B., Carrol, F. S., & Garcia, B. (1970).   Temporal order in the perception of vowels. *Journal of the Acoustical Society of America*, 48, 1010-1013.

21   Analysis of a Simultaneous-Speaker Sound Corpus

Brian L. Karlsen, Guy J. Brown, Martin Cooke, Malcolm Crawford, Phil Green and Steve Renals
University of Sheffield

In this chapter we present the results of an analysis of the ShATR simultaneous-speaker corpus, which is primarily intended to provide acoustic material for studies in auditory scene analysis. Our analysis includes transcription and alignment at four different levels, overlap analysis between speakers, and word counts. We also describe a general tool for accessing concurrent events in transcribed multi-sound-source databases.

21.1 INTRODUCTION

Spoken communication usually takes place in an acoustically cluttered environment. There are typically several sound sources present, whose number and characteristics cannot always be predetermined. The Sheffield-ATR simultaneous speaker database (Crawford et al., 1994) (ShATR) is a new corpus that was collected to facilitate research on speech perception in such natural surroundings. The recordings were made at the Advanced Telecommunications Research Institute International (ATR), Kyoto. The purpose of this chapter is to report on the analysis of these data.

Human listeners have a remarkable ability to pay selective attention to individual sound sources, a feat referred to as *auditory scene analysis* (ASA) (Bregman, 1990). Work at Sheffield (Cooke, 1993; Brown & Cooke, 1994) has achieved some success in computational modeling of ASA based on primitive grouping principles such as common onset, periodicity, and good continuation. We have also started to address the problem of how ASA might be used in the task of automatic speech recognition (ASR) in noise (Cooke et al., 1994; Green et al., 1995).

Computational ASA research has now reached the point where a corpus of auditory scenes is required for training and evaluation of segregation and recognition algorithms. Most existing speech corpora (e.g. TIMIT (Garofolo & Pallett, 1989); RM (Price et al., 1989); WSJ (Paul & Baker, 1992)) are unsuitable because they consist solely of clean single-channel speech. The NOISEX database (Varga et al., 1992) provides speech with added noise of various types and at various signal-to-noise ratios. This material, however, is not typical of auditory scenes. The noise and speech are recorded separately, producing no Lombard effect (Summers et al., 1988).

Our aim was to record a multiple sound-source corpus with the following requirements:

(1) There should be several speakers;

(2) The speakers should engage in a collaborative task;

(3) The task should be sufficiently natural to provoke spontaneous speech;

(4) The execution of the task should generate additional non-speech sounds;

(5) For a significant proportion of the time more than one sound source should be active;

(6) The task should provide suitable test material for large-vocabulary speech recognition and common keyword-spotting.

The task we devised to meet these requirements was based on solving crossword puzzles. The task had an unrestricted vocabulary but induced frequent occurrences of a few words, such as *across*, *down*, *blank*, the numerals from 1 to 30, and so on, hence ensuring requirement (6) above. Pilot studies led us to adopt the following arrangement:

• Five speakers were seated around a table (1);

• There were two teams of two speakers, each team solving a different crossword puzzle (2). In single conversations, overlapping speech is uncommon (Sacks et al., 1974). Having two teams allows us to fulfill requirement (5) above;

• The fifth speaker had the answers to both crosswords and was allowed to give hints;

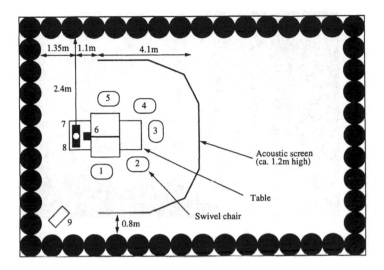

FIG. 21.1. Plan of the recording chamber (not to scale) and equipment set up.

- Seating was arranged so that it was necessary for each team to "talk across" the other team, and communicate with the hint-giver.

A diagram showing the layout of the recording environment is given in Figure 21.1, and a photograph is shown in Figure 21.2. The hint-giver took position 3, and the participants were paired into teams (1, 4) and (2, 5). This served to create a more interesting acoustic environment than (1, 2) and (4, 5) pairings, because more speech was directed across the mannikin; members of each team could not "huddle" together.

We found that the task was sufficiently absorbing, and sufficiently collaborative, to fulfill (3) above, and that there were abundant nonspeech noises such as moving chairs, paper rustle and so on to satisfy (4). Our analysis of the degree of overlap between speakers confirmed that (5) was fulfilled (see Figure 21.5).

Prior to taking part in the task, each subject went through an enrollment procedure to provide data that could be used for training, or adapting, speech or speaker recognizers. This consisted of reading the following while sitting in the same location in which the speaker would be during the task:

(1) The TIMIT SA sentences (Garofolo & Pallett, 1989), twice each;

(2) Ten repetitions each of the words *across*, *down*, *yes*, and *no*, interspersed in random order;

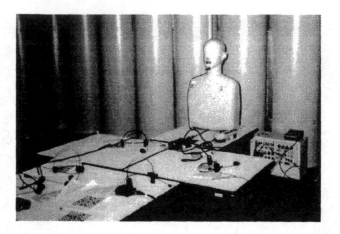

FIG. 21.2. Photograph of the recording chamber.

(3) Ten repetitions each of the letters of the alphabet and digits from 0 to 30 (in random order);

(4) A passage in English from *The Japan Times*.

Although several sessions were recorded, this chapter reports on the analysis of a single session, which lasted 37 minutes and which was recorded with 48 kHz sampling rate, during which the two teams of crossword solvers solved one cryptic crossword each with the help of the hint giver.

21.2 ANALYSIS

The recordings were analyzed at four different levels:

(1) The overall structural level: people coming into the room, sitting down, putting on microphones, etc.

(2) The sound type level: whether it was a particular nonspeech event, like *uh*, *um*, *ooh*, paper-rustle and so on, or a speech event.

(3) The orthographic level: both singular words and composite "sentences".

(4) The phone level.

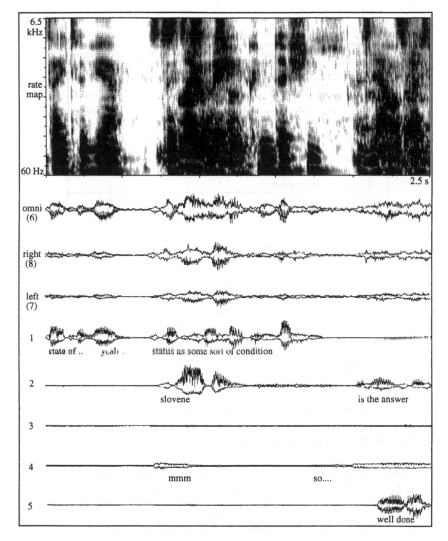

FIG. 21.3. A cross-section of the 8 channels at a particular interval in time with an auditory rate representation of the omnidirectional channel shown at the top.

The analysis produced an alignment of all the described events at every level. The degree of overlap between the individual speaker channels at the orthographic level was also analyzed. In Figure 21.3 one can see a typical cross section over the 8 recorded channels.

Table 21.1 The distribution of structural level categories.

Category	% of total time
silence	0.1
coming into the room	0.9
sitting down	0.1
putting on microphones	1.4
discussion before the task	6.0
doing the crossword task	88.6
discussion after the task	0.8
taking off microphones	0.1
leaving the room	1.9

21.2.1 Structural level analysis

The analysis of the overall structural level was carried out by listening to the recordings obtained with the omnidirectional microphone. In this way it was possible to identify shifts of context which acted as the basis for the categorization. The different categories found are the following:

> Silence, people coming into the room, sitting down, putting on microphones, discussing before the task, doing the crossword task, discussing after the task, taking off microphones, and leaving the room.

Distribution of these categories is given in Table 21.1.

21.2.2 Sound level analysis

The sound level analysis involved labeling acoustic events on individual channels. For this purpose we automatically extracted segments of the sound files that contained statistically high activity in relation to the background level. These segments were manually transcribed using an extended set of labels previously used for the WSJCAM0 corpus (Fransen et al., 1994):

> speech, speech-plus, whisper, *er, um, uh*, paper-rustle, chair-squeak, door-slam, mic-noise, *oh, ooh*, lipsmack, loud-breath, cross-talk, laughter, unintelligible, cough, *mm*, throat-clear, grunt, tongue-click, sigh, *ah*, tap.

The distribution of nonspeech and speech sounds in relation to total recording time for each individual speaker can be found in Figure 21.4. Speaker 2 is the most active by far. However, this is balanced by the lesser activity of speaker 2's crossword partner, speaker 5, so that the two teams (1,4 and 2,5) contribute

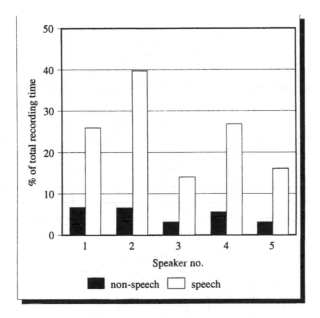

FIG. 21.4. Nonspeech and speech in relation to total recording time.

approximately equal amounts of material. It is also clear from this analysis that the hint giver (speaker 3) contributes least to the conversation.

21.2.3 Orthographic level analysis

The segments categorized as "speech" and "speech-plus" (i.e., an interval containing speech and one of the other categories) were extracted for further analysis at the orthographic level. These segments correspond to the aforementioned composite "sentences". Word alignment was performed automatically as described below. Such segments were manually marked and orthographically transcribed. An overlap analysis between the individual speaker channels was carried out, the results of which are shown in Figure. 21.5. This analysis reveals that two or more participants spoke simultaneously approximately 38% of the time.

Word counts were also performed producing the following results: A total of 8313 words were spoken in this session. Individual speakers contributed between 2950 and 997. Table 21.2 gives counts, by speaker, for content words common in the crossword domain. Table 21.3 shows counts for the digits 1 to 9.

Table 21.2: Counts of common content words.

Word	Speaker 1	2	3	4	5	Total
across	18	24	4	16	8	70
anagram	9	5	4	7	7	32
blank	0	19	0	8	1	28
down	9	38	7	23	12	89
hint	1	10	4	4	4	23
letters	4	9	1	7	1	22
right	18	12	14	9	6	59
word	7	9	8	9	2	35

Table 21.3: Digit word counts.

Word	Speaker 1	2	3	4	5	Total
one	26	20	6	27	13	92
two	17	12	6	11	4	50
three	16	7	2	6	3	34
four	17	10	4	10	4	45
five	11	3	1	10	2	27
six	11	6	1	5	3	26
seven	8	7	9	19	5	48
eight	4	6	8	4	4	26
nine	5	4	1	6	2	18

21.2.4 Word alignment and phone-level analysis

Phone and word alignments were performed simultaneously using the British English version of the ABBOT large vocabulary speech recognition system (Robinson et al., 1995). This system is a hybrid connectionist/HMM system that uses a recurrent network to estimate phone probabilities.

ABBOT uses the British English Example Pronunciation (BEEP) dictionary,[1] which contains British English pronunciations for around 250,000 words. The phone set used in BEEP is an adaptation of the ARPAbet symbols used in the TIMIT database (with additions to account for British English vowel space).

Since BEEP was primarily derived from textual sources, it was necessary to augment it with words that are characteristic of spoken language (e.g., the contraction *what've*). This was achieved by asking native British English

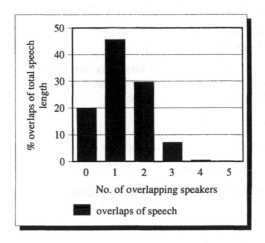

FIG. 21.5. Speech overlap analysis.

speakers, familiar with the phone notation of BEEP, to transcribe the spoken words independently, so that each word would be transcribed at least two times by two different speakers. The two transcriptions were then compared: If transcriptions differed but were both reasonable, each was included in the dictionary.

Orthographic alignment and phone level transcription using the ABBOT recognizer is an automatic process. An additional advantage of using a statistically based recognizer is that the alignment probabilities may be used to identify possibly erroneous transcriptions (a low alignment probability may be an indication of a transcription error).

21.3 THE SPEECH DATABASE FINDSEGS TOOL

The usefulness of a labeled corpus is to some degree dependent on ease of access to segments conforming to a desired specification. In ShATR, whose focus is simultaneous acoustic material, it is essential to provide a general purpose segment extraction tool which allows the specification of temporally concurrent patterns. The findsegs tool addresses this requirement.

The findsegs tool applies an algorithm which is capable of searching a number of different files in parallel to extract segments that correspond to a user's query. The query language has the following features:

- User-defined terminals and nonterminals: For example,

  Voiced=ahlaalawl...
  Fricated=flslzl...

- Concurrent patterns in different channels overlapping with each other: For instance, to find all voiced segments overlapping with fricatives in other channels, the query would simply be

  Voiced AND Fricated

- Left and right contexts for segments:

  Voiced)Fricated(ah

(Note the free mixing of terminals and nonterminals.)

- Alternatives:

  VoicedlFricated

Note that this query is addressing alternatives within one channel (either Voiced or Fricated), as the AND mechanism above is the only means of specifying concurrent patterns.

- Sequences:

  {Voiced,Cons,Voiced}

- Negations, conjunctions and disjunctions of nonterminals:

  !((Voiced&Fricated))

In words: not voiced and fricated. Note that the & in this query is a logical operation, and it should not be confused with the AND previously mentioned, which indicates overlaps.

These can be combined to build complex queries such as:

  Cons)Voiced(Cons AND (Voiced&Fricated))

which will return just the voiced regions of CVC syllables which overlap with voiced fricatives. All pairs of channels will be searched. Additionally the user can specify a subset of channels to be searched, and must define the level of analysis to which the query is directed.

Findsegs can be used with traditional single channel databases such as TIMIT. The tool operates in both command line mode and in an interactive mode based on a standard VT100 style terminal, and is implemented in ANSI C for portability across a range of platforms. This tool has now been released on the ShATR CDROM.[2]

21.4 CONCLUSIONS

We have described the analysis of a multi-source, multi-sensor corpus primarily designed for auditory scene analysis research. The corpus contains substantial amounts of overlapping acoustic material, annotated at various levels with easy access to specified patterns. The corpus can be used for localization studies and for the assessment of monaural and binaural segregation schemes. In addition, although the amount of data is insufficient for training automatic speech recognition systems, the corpus provides a challenging domain for word spotting, speaker identification, and large-vocabulary spontaneous speech recognition systems. Other application domains include the analysis of prosody, intention, dialogue and discourse.

The ShATR corpus and associated tools and transcriptions have now been released on CDROM. In addition, a sample from the database, corresponding to the mannikin channels of the segment displayed in Figure 21.3 is available via FTP and WWW.[3]

21.5 ACKNOWLEDGMENTS

We would like to thank Dr. M. Tsuzaki, Dr. H. Kawahara, and colleagues in the Human Information Processing Laboratory at ATR for their cooperation in obtaining the recordings, David Kirby at the BBC for his help in the early stages of configuring the recording equipment, and Inge-Marie Eigsti for her willing participation as a subject. This research is supported by an EPSRC Standard Research Grant GR/K22638.

REFERENCES

Bregman, A. S. (1990). *Auditory Scene Analysis*. MA: MIT Press.

Brown, G. J. & Cooke, M. P. (1994). Computational auditory scene analysis. *Computer Speech and Language*, 8, 297-336.

Cooke, M. P. (1993). *Modelling Auditory Processing and Organisation*. UK: Cambridge University.

Cooke, M. P., Green, P. D. & Crawford, M. D. (1994). Handling missing data in speech recognition. *Proceedings of the International Conference on Speech and Language Processing*, pp. 1555-1558. Yokohama.

Crawford, M. D., Brown, G. J., Cooke, M. P. & Green, P. D. (1994). Design, collection and analysis of a multi-simultaneous-speaker corpus. *Proceedings of the Institute of Acoustics*, 16, 183-190.

Fransen, J., Pye, D., Robinson, T., Woodland, P. & Young S. (1994). WSJCAM0 corpus and recording description. Technical report no. CUED/F-INFENG/TR.192. UK: Department of Engineering, University of Cambridge.

Green, P. D., Cooke, M. P. & Crawford, M. D. (1995). Auditory scene analysis and HMM-recognition of speech in noise. *Proceedings of the IEEE International Conference on Acoustics, Speech and Signal Processing (ICASSP-95)*, pp. 401-404.

Garofolo, J. S. & Pallett, D. S. (1989). Use of the CD-ROM for speech database storage and exchange. *Proceedings of the European Conference on Speech Communication and Technology*, pp. 309-315. Paris.

Paul, D. B. & Baker, J. M. (1992). The design for the wall street journal-based CSR corpus. *Proceedings of the International Conference on Speech and Language Processing (ICSLP-92)*, pp. 899-902. Banff.

Price, P., Fisher, W. M., Bernstein, J. & Pallet, D. S. (1988). The DARPA 1000-Word resource management database for continuous speech recognition. *Proceedings of the IEEE International Conference on Acoustics, Speech, and Signal Processing (ICASSP-88)*, P. 651-654.

Robinson, T., Fransen, J., Pye, D., Foote, J. & Renals, S. (1995). WSJCAM0: A British English speech corpus for large vocabulary continuous speech recognition. *Proceedings of the IEEE International Conference on Acoustics, Speech, and Signal Processing (ICASSP-95)*, pp. 81-84. Detroit.

Sacks, H., Schegloff, E. A. & Jefferson, G. A. (1974). A simplest systematics for the organization of turn-taking for conversation. *Language*, 50, 697-735.

Summers, W., Pisoni, D., Bernacki, R., Pedlow, R. & Stokes, M. (1988). Effects of noise on speech production: Acoustic and perceptual analyses. *Journal of the Acoustical Society of America*, 84 (3), 917-928.

Varga, A., Steeneken, H. J. M., Tomlinson, M. J. & Jones, D. (1992). The NOISEX-92 study on the effect of additive noise on automatic speech recognition (CD-ROM). Malvern, UK: DRA, Speech Research Unit.

NOTES

[1] Available via internet FTP from
ftp://svr-ftp.eng.cam.ac.uk/pub/comp.speech/dictionaries/beep-1.0.tar.gz .

[2] It is also available via FTP from
ftp://ftp.dcs.shef.ac.uk/share/spandh/ShATR/findsegs-1.0.tar.gz .

[3] See
http://www.dcs.shef.ac.uk/research/groups/spandh/pr/ShATR/ShATR.html.

22

Hearing Voice: Transformed Auditory Feedback Effects on Voice Pitch Control

Hideki Kawahara
ATR Human Information Processing Research Laboratories

Real-time interaction between speech perception and production is quantitatively revealed by a new measurement technique called TAF (Transformed Auditory Feedback) for voice pitch control. Experimental results suggest that there are two major auditory processes which control voice pitch and that they work in parallel to produce a pitch control signal. One response has a relatively fast (4 Hz to 7 Hz, typically) natural frequency and a delay of 90 ms to 120 ms depending on the pitch. The other response has a relatively slow (0.5 Hz to 1 Hz, typically) natural frequency and a maximum effect around 300 ms to 600 ms. They are both compensatory responses. Implications for the information processing architecture of speech communication mechanisms and the biological basis will also be discussed.

22.1 INTRODUCTION

The basic questions addressed in this chapter are as follows: (1) Does speech perception play an indispensable role in speech production? (2) Are there real-time interactions between speech perception and production? (3) If such interactions exist, how do they interact? We will answer these questions by providing a new procedure to enable quantitative measurement of such interactions. In short, the answer to the first two questions is "Yes," and the results shown in this chapter answer the third question. First of all, a brief introduction provides the background.

It is generally believed that we monitor our own voice while speaking. If this is the case, speaking will be affected by auditory stimulation, which alters natural feedback conditions. Reactions to DAF (Delayed Auditory Feedback) and the Lombard effect are typical examples. Speaking under DAF conditions with about 200 ms delays tends to produce effects like stuttering and is very disturbing (Lee, 1950). Instinctive modification of speaking style and physical characteristics under noisy environments (Pisoni et al., 1985), which are common in the Lombard effect, are found to produce difficulties in automatic speech recognition. It was also reported that there are low-level laryngeal reflexes to auditory stimulations, such as loud clicks and frequency modulated tones (Sapir et al., 1983a; 1983b).

These effects suggest that the auditory system plays an important role even in speech production. It may indicate that component processes of speech communications are interactive in nature like other human behaviors. If this is the case, it is indispensable that we understand these interactions in order to model our natural speech communication functions. However, there have been few quantitative investigations on interactions between speech perception and production under normal speaking conditions. Until recently, virtually nothing was known about the time course of these interactions. TAF was developed to analyze these interactions without major interference to normal speech communication processes (Kawahara, 1993a; 1993b). It was substantiated, in my previous report (Kawahara, 1994), that the interaction estimated using TAF actually exists in voicing without artificial feedback or perturbations, even though the report could not provide a reliable transfer function from perception to production.

In this chapter, we will present quantitative examples of interactions between auditory perception and speech production estimated by TAF analysis procedures in voice pitch control. These results also imply that we have obtained a unique and powerful procedure to estimate auditory functions objectively in a noninvasive manner.

22.2 BASIC IDEA OF TAF

The basic assumptions underlying the TAF method are as follows. (1) If interactions mediated by auditory processes exist, the perturbation of speech parameters in the auditory feedback path for a speaker's own voice will modify the speech parameters of the produced voice. (2) If the perturbation is small enough, the interaction will be approximated by the behavior of a linear system. (3) There is a condition where the perturbation is small enough not to disturb normal speech production behavior but large enough for its effects on speech parameters to be detectable. (4) If the interactions consist of multiple modules, response characteristics can be decomposed. The last condition is interesting for understanding internal processes, but it is not crucial for TAF to be possible.

TAF is a general method to detect the effects of auditory feedback using parametric perturbations which are inserted into an auditory feedback path. Voice pitch was selected as the parameter to be modulated for the current experiments, although any speech parameters can be used as the parameters to be modulated, in principle.

There were several reasons to believe that the largest effect of parametric perturbations would be observed for pitch modifications. For example, (1) voice pitch is unstable in nature without feedback, (2) it was reported that DAF conditions without phonological information feedback also produce similar DAF effects (Howell & Archer, 1984), and (3) it was also reported that prosodic information is the first speech information acquired by infants (Fernald & Kuhl, 1987).

The final point may need additional explanation. It is based on an assumption that the human information processing structure for speech communications reflects the developmental process of language acquisition. If this holds, then it is likely that a skill which is acquired in the very early stages of this developmental process may make use of low-level real time control. In other words, higher skills, like the production of consonants, may not use real time feedback control based on auditory information, because the large amount of delay caused by the consonant perception process yields output well after the completion of the consonant production process.

22.3 IMPLEMENTATION OF TAF FOR PITCH INTERACTIONS

22.3.1 Requirements

The perturbation signal has to be small enough not to disturb natural speech production but large enough for the effects to be detected. This condition requires the signal to provide a systematic way of improving the signal-to-noise ratio, and requires the maximum perturbation level to be minimal for a given total perturbation energy. It is also desirable for the signal to consist of many frequency components, and for the perturbation signal to be complex enough not to be easily predicted by subjects.

The other condition for the perturbation signal concerns the effective frequency range. The effective range should be well below the fundamental frequency of the target speech, because the periodic excitation of voice sounds has an effect like the sampling pulses of a discrete time system.

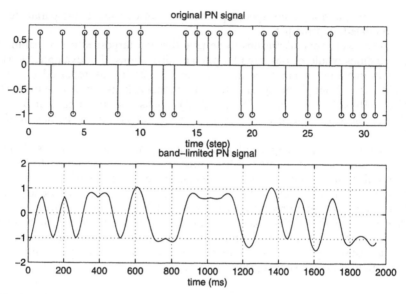

FIG. 22.1. The employed perturbation signal. The upper plot shows the original 31-step PN signal. The lower plot shows the band-limited PN signal, which was used throughout the experiments. These plots represent one cycle.

22.3.2 Method

A pseudorandom noise (a PN signal) (Miyakawa, 1978) based on the M-sequence with band limitation was used as the perturbation signal, because it fulfills the requirements and listed above and because translated PN signals are orthogonal to each other. Periodic averaging was introduced to reduce the effect of random errors, using the periodic nature of the PN signal. The original M-sequence which was used to produce the PN signal consisted of 31 independent elements (cf. appendix). The PN signal was over-sampled and smoothed to produce the band-limited perturbation signal. The repetition period of the perturbation signal was 1.9375 seconds Figure 22.1 illustrates the original PN signal and the band-limited PN signal.

Figure 22.2 shows a diagram of the TAF experiment for pitch perturbation. The speaker's voice, detected by a microphone, was modulated by a pitch converter according to the perturbation signal. The modulation was multiplicative in relation to the fundamental frequency. In other words, perturbation is additive in the frequency domain. (There are no practical differences between multiplication and addition, because the amount of perturbation is of the average fundamental frequency.) The modulated voice was fed back via circumaural headphones to the speaker. In this experiment, the feedback loop gain was set 15dB to 20dB higher than the natural sidetone via air

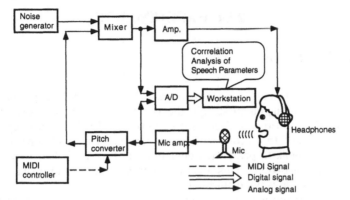

FIG. 22.2.   A schematic diagram of how measurements are made with Transformed Auditory Feedback.

conduction. In addition to this excess gain, a pink noise of 78dB(A) was mixed with the fed-back voice to mask the natural sidetone. Both the fed-back voice pitch and the produced voice pitch were extracted using the same pitch extraction algorithm (Secrest & Doddington, 1983) to cancel possible artifacts that might be introduced by the pitch converter.

Pitch trajectories were analyzed in the following steps:

(1) Synchronized averaging using periodicity,

(2) Periodic cross-correlation computation to estimate the time compressed perturbation signal and the approximated closed loop impulse response,

(3) Alignment of the phase of the estimated signals,

(4) Integration of signals estimated by separate measurements,

(5) Estimation of the coherency,

(6) Estimation of the open-loop transfer function,

(7) Estimation of the open-loop impulse response,

(8) Decomposition of the open-loop impulse response using second order models,

(9) Validation of the decomposition results by re-synthesis of the transfer function.

22.4 EXPERIMENTS

22.4.1 Conditions

Six subjects, 3 male and 3 female speakers, participated in the experiments. The subjects were instructed to sustain the vowel /a/ for about 8 seconds under several TAF conditions and control conditions. Each subject repeated the sustained voicing 5 times for each pitch condition. The conditions are given in Table 22.1. A peak-to-peak pitch deviation of 50 cents (a half semitone) was used as the perturbation. This level of pitch deviation is equivalent to natural pitch deviations and is not disturbing.

A following series of experiments using a sustained vowel condition was conducted. Four male subjects and two female subjects participated in this series. The subjects were asked to voice the sustained vowel /a/ for about 10 seconds or more repeatedly, during each 2-minute session. One recording was made for one target pitch condition for each subject.

In addition to these sustained vowel experiments, one subject repeatedly read one sentence for 46 minutes under the TAF condition. The Roman transcription of the sentence is "aioi no oi wa, yama no ue no ie ni iru," which consists of only voiced sounds.

Table 22.1. List of target frequencies for pitch control.

Note ID	male (Hz)	female (Hz)
G2#	103.8	207.6
C3	130.8	261.6
E3	164.8	329.6
G3	207.7	541.3

22.4.2 A Tour

In this section, we describe in detail the steps used to analyze exemplar data obtained in one 2-minute session. The subject was a male, and the target pitch was G3# (207.7 Hz).

Approximated Closed-Loop Impulse Response. Figure 22.3 shows a cross correlation with a band-limited PN signal and cyclic averaged fundamental frequency trajectories for the fed-back voice and produced voice. The lower plot, which is the cross correlation for the produced voice, approximates a closed-loop impulse response, because perturbation was added to the produced speech. The dash-dot lines in the lower plot indicate a confidence interval of 90 reliability. It is clearly illustrated that the fundamental frequency of the produced speech responds in a compensatory manner to the perturbation with a

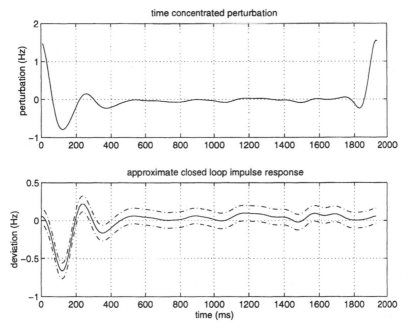

FIG. 22.3. An exemplar perturbation and an approximate closed loop impulse response. The dash--dot lines in the lower plot show a 90 confidence interval.

latency of about 130 ms. The confidence interval presented in the same figure suggests that the compensatory response is not an artifact.

Estimated Open-Loop Transfer Function.   Figure 22.4 shows results of a transfer function analysis of the same data as Fig. 22.3. The transfer function was estimated as the ratio of Fourier transformation of the averaged response and the averaged perturbation. The top plot shows coherency, which was calculated by the estimated error variance, and indicates the reliability of estimation at each frequency. The second plot illustrates the open loop gain for the perception to production transfer function. The bottom plot represents the open loop phase shift of the transfer function. Again, the dash--dot lines represent confidence intervals.

One remarkable feature, which was not found in our previous experiments, is that the gain in the low frequency region (1 Hz or less) is almost unity or more. This suggests that with a proper phase shift, the voice pitch control system can easily be unstable. This is exactly what happened when a 500 ms delay was inserted into the artificial feedback path. In the higher frequency region, the absolute gain is around -8 dB and is relatively flat. The corresponding phase plot is roughly linear in the 2 Hz to 6 Hz region. These features suggest that the compensatory response found in our previous experiments was caused by the transfer function in this frequency range.

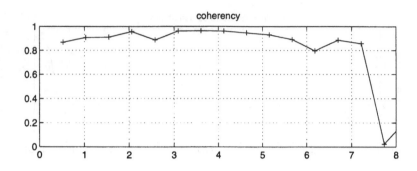

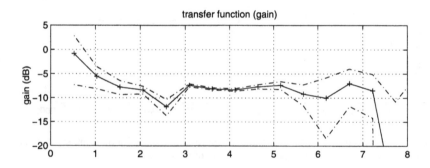

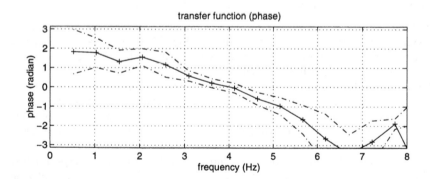

FIG. 22.4. An example of the estimated loop transfer function and coherency for a male subject. The dash--dot line shows a 90 confidence interval.

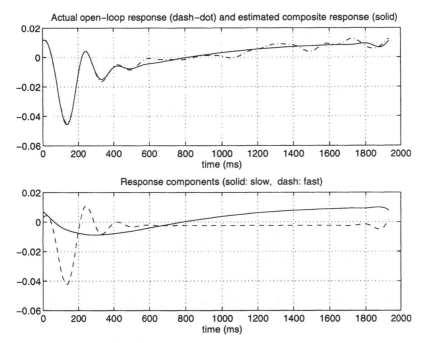

FIG. 22.5. An example of response decomposition. The dashed line in the top plot represents the original open loop impulse response. The solid line in the same plot represents the composite response using the two second-order responses shown in the bottom plot.

4.2.3 Decomposition of Open-Loop Impulse Response. Figure 22.5 shows a decomposition result. The open loop impulse response (dash–dot line in the upper plot), which was estimated by the inverse Fourier transformation of the band-limited estimated transfer function, is closely approximated by a composite signal (solid line in the upper plot) of two second-order responses. The component responses are represented in the lower plot of the same figure. This decomposition was performed by a constrained nonlinear parameter optimization using minimum squared error criterion in the time domain. Gaussian error weighting in the frequency domain was also introduced. The response model and parameters to be optimized were listed in the appendix.

One important observation in this result is that the relatively fast component consists of a delay of about 100 ms. In other words, the fast second-order response starts 100 ms after stimulation. This response time may reflect the information processing performed to compute neural commands from the feedback voice. It consists of time for pitch extraction, error detection, and motor command generation.

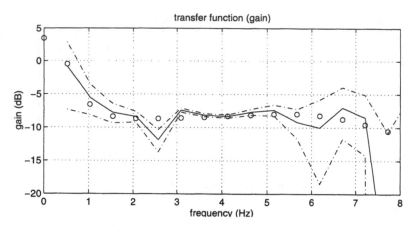

FIG. 22.6. Resynthesized open loop transfer function. The open circles represent the re-synthesized characteristics. The solid line and dot-dash lines represent the estimated open loop transfer function and its confidence interval.

In this example, the natural frequency for the fast response was 6.6 Hz, and its damping factor was 0.37. The natural frequency for the slow response was 0.5 Hz, and its damping factor was 1.

4.2.4 Resynthesized Transfer Function. Finally, Fig. 22.6 shows the resynthesized open loop transfer function (open circles in the figure) and its original counterpart (solid line in the figure). Nearly all the resynthesized transfer function gains are within the confidence interval of estimation (dash--dot line in the figure). The figure illustrates that the response was closely approximated by the composite response also in the frequency domain.

22.4.3 Results

Two groups of data were analyzed, according to the steps described in the previous section, and pooled for statistical analysis. The total amount of data were 13 (8 male and 5 female subjects) times 4 different frequencies.

Experimental results obtained using a sustained vowel showed that the response to the perturbation consisted of two components, as demonstrated in the tour. These two component responses were integrated by simple addition in log-frequency representation. One response had a relatively fast (4 Hz to 7 Hz, typically) natural frequency and a delay of 90 ms to 120 ms depending on pitch. The other response had a relatively slow (0.5 Hz to 1 Hz, typically) natural frequency and a maximum effect around 300 ms to 600 ms. They were both compensatory responses.

Figure 22.7 shows the averaged delay versus the pitch of the produced voice. The vertical bars represent standard deviations among subjects. The dashed line represents a hypothetical delay that consists of a constant processing

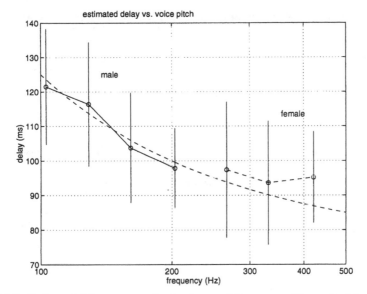

FIG. 22.7. Estimated delay versus voice fundamental frequency. The vertical bars in the plot represent standard deviations. The dashed line in the figure is a hypothetical information processing duration.

time for error detection and motor command generation, and pitch dependent processing time for pitch extraction. The dashed line assumes that the pitch extraction process needs 5 repetitions of a fundamental period of the speech waveform. This will be discussed again, later.

Figure 22.8 shows a decomposition result for the repeated sentence. The response found was similar to that found in the case of the sustained vowel. It was also decomposed into two components with different time constants. The natural frequency of these responses resided within the normal range of those of sustained vowels. However, the confidence interval of the approximated open-loop response indicated that more than 3 hours of data is necessary to provide similar reliability.

22.5 DISCUSSION

These results clearly provide an answer to the third question by illustrating impulse responses and their decompositions. The interaction is in real time. This provides a "Yes" answer to the second question asked at the beginning of the chapter. The high gain region in the low frequency range is strong but indirect evidence that auditory feedback plays an important role in controlling

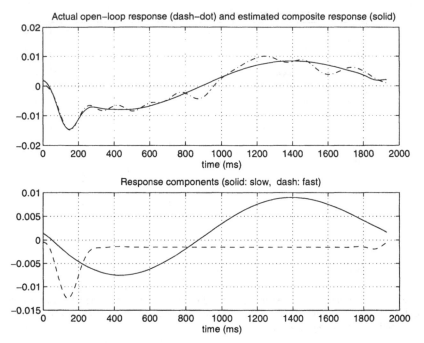

FIG. 22.8. Response decomposition in the case of a repeated sentence. The sentence only consists of voiced sounds. The dashed line in the top plot represents the original open loop impulse response. The solid line in the same plot represents the composite response using the two second-order responses shown in the bottom plot.

pitch in speech production. The similar response found in the read speech case supports this possibility. In addition to this evidence, taking into account the results given in our previous report (Kawahara, 1994), it is safe to state that speech perception does play an indispensable role in speech production. In other words, the answer to the first question is also "Yes." In addition to answering these questions, the TAF results also open up interesting questions to be investigated.

First, this may provide a new quantitative tool to investigate our pitch perception mechanism. The dependencies of the delay of the fast component for male and female subjects seemed to follow the same trend. This may suggest that the trend is due to a common mechanism that is dependent not on gender but on pitch itself. The most possible candidate is the amount of time for pitch extraction based on temporal processing. This is because the delay of the fast component has to represent the total information processing time required to generate the pitch control signal from the observed speech waveform. Even though the dashed line based on the temporal pitch extraction assumption corresponds with the measurements within the observation errors, it seems like a bending exists around 250 Hz. This is an interesting coincidence with the transition frequency of two different (temporal versus spatial) pitch perception mechanisms (Pierce, 1991).

The additive integration of component responses is also interesting because it suggests that higher information processing in the human brain does not subsume lower activities completely. The preliminary experiment using read text also supports this observation. Two auditory-mediated feedback systems working in parallel provoke the speculation that the fast response primarily represents autonomous control via the cerebellum and that the slow response primarily represents higher cognitive functions. It will be interesting to see whether the slow response vanishes in more natural speech conditions, including spontaneous speech. That will provide a clue to investigating how speech communication activities are programmed in our brain and what type of unit is used to control speech production processes. Another important open question is how to correlate these results with a comprehensive model of the fundamental frequency contour generation process (Fujisaki, 1992).

22.6 CONCLUSION

In this chapter, I have demonstrated that a real-time interaction exists between speech perception and production in pitch control, by using a new measurement procedure called TAF. The interaction is basically a compensatory response to pitch fluctuations. One interesting finding is that the response can be decomposed into two second-order responses with different time constants. Discussions about the nature of these two responses and their relation to information processing architectures in speech communications are also given.

These findings provide quantitative evidence showing that our voice is actually "hearing" itself and other environmental sounds while we speak. They also indicate that the TAF method can be a powerful tool in investigating human speech communication mechanisms.

ACKNOWLEDGEMENT

The author appreciates the work of Ms. J. C. Williams of Ohio State University for her discussions and help in the data acquisition. He also wishes to thank Dr. Alain de Cheveigne and Mr. James Magnuson for their comments.

REFERENCES

Fernald, A. & Kuhl, P. K. (1987). Acoustic determinants of infant preference for mother's speech. *Infant Behavior and Development*, 10, 279-293.

Fujisaki, H. (1992). Modeling the process of fundamental frequency contour generation. In Tohkura, Y., Vatikiotis-Bateson, E., Sagisaka, Y. *Speech Perception, Production and Linguistic Structure* , pp. 313-326. IOS Press.

Howell, P. & Archer, A. (1984). Susceptibility to the effects of delayed auditory feedback. *Perception and Psychophysics*, 36 (3), 296-302.

Kawahara, H. (1993a). On interactions between speech production and perception using transformed auditory feedback H-93-24, Acoust. Soc. Jpn. Hearing TC. (in Japanese).

Kawahara, H. (1993b). Transformed auditory feedback: Effects of fundamental frequency perturbation. *J. Acoust. Soc. Am.*, 94 (3 Pt2), 1883.

Kawahara, H. (1994). Effects of natural auditory feedback on fundamental frequency control. In *Proc. ICSLP'94*, 1399-1402, Yokohama.

Lee, B. S. (1950). Effects of delayed speech feedback. *J. Acoust. Soc. Am.*, 22 (6), 824-826.

Miyakawa, H. (1978). *Probabilistic Systems and Estimation of Dynamic Characteristics*. Colona. (in Japanese).

Pierce, J. R. (1991). Periodicity and pitch perception. *J. Acoust. Soc. Am.*, 90 (4), 1889-1893.

Pisoni, D., Beranacki, R. H., Nusbaum, H. C., Yuchtman, M. (1985). Some acoustic phonetic correlates of speech production in noise. *Proc. IEEE ICASSP85*, pp. 1581-1584.

Sapir, S., McClean, M. D., & Larson, C. R. (1983a). Human laryngeal response to auditory stimulation. *J. Acoust. Soc. Am.*, 73 (1), 315-321.

Sapir, S., McClean, M. D., & Luschei, E. S. (1983b). Effects of frequency-modulated auditory tones on the voice fundamental frequency in humans. *J. Acoust. Soc. Am.*, 73 (3), 1070-1073.

Secrest, B. G. & Doddington, G. R. (1983). An integrated pitch tracking algorithm for speech systems. *Proc. IEEE ICASSP83*, pp. 1352-1355.

APPENDIX

A PN signal

The original PN signal $s(n)$ was generated using the following recursive equation.

$$s(n) = X_n \left(1 \pm 2^{-\frac{q}{2}} \right) - 1 \qquad \text{EQ 22.1}$$

$$X_n = X_{n-p} \oplus X_{n-q} \qquad \text{EQ 22.2}$$

Here $\oplus$ denotes an 'exclusive OR' operation, and $p < q$. The parameters, $q = 5, p = 2$, were selected for the current experiment.

B Optimization parameters for the composite 2nd order response model

Let's assume that $h(t)$, the open-loop transfer function estimated by the TAF procedure, is a summation of two 2nd order responses, $r_f(t) = \tilde{g}_f(t - \tau_f)$ and $r_s(t) = \tilde{g}_s(t - \tau_s)$. Where g_f or g_s is an impulse response of a 2nd order system $g(t)$ which is characterized by the damping factor ζ and the natural frequency ω_n. Parameters τ_f and τ_s represent delay for the response to start. The sign ~ represents periodic signal, because every operation is performed on periodic signals due to the periodicity of the PN signal.

$$g(t) = \frac{\omega_n}{\sqrt{1 - \zeta^2}} e^{-\zeta \omega_n t} \sin \sqrt{1 - \zeta^2} \, \omega_n t \qquad \text{EQ 22.3}$$

Where $g(t) = 0$ for $t < 0$
The form of $g(t)$ is in the case of $\zeta < 1$. The parameters to be optimized are:
$$\theta = \{a_f, a_s, a_0, \omega_f, \zeta_f, \tau_f \omega_s, \zeta_s, \tau_s\}.$$
The parameters a_f and a_s represent the amplitude of each response and a_0 represents a bias.

23

Toward Content-Based Audio Indexing and Retrieval and a New Speaker Discrimination Technique

Lonce Wyse and Stephen W. Smoliar
National University of Singapore

Several techniques for identifying segment transitions in an audio stream are discussed. Gross features are first identified that control more detailed and computationally expensive analysis down stream. Pitch is tracked using some basic streaming principles, and then used as one cue to speaker transitions. A novel speaker discrimination technique is described that makes segmentation decisions when a continuously updated model of the current speaker suddenly ceases to sufficiently account for the input data.

23.1 INTRODUCTION

Despite the multimedia hype, video and audio information are not currently part of our everyday computing environment. We do not yet have the tools for manipulating this kind of information with the ease with which we manipulate text. The goal of the Video Classification Project at ISS is to automatically segment a stream of image and sound data into meaningful units which can then be used in a database system (Smoliar and Zhang, 1994). We consider the problems of parsing input streams, automatic indexing (labeling) of segments, and retrieval techniques. Such a system will support nonlinear browsing of material and the use of sound and image keys for retrieval, which are far more natural ways of interacting with multimedia data than simple linear scanning.

Currently, the audio and video stream parsing is done separately. However, the new systems will run together, because each separate media stream contains information that can help the other make parsing decisions. The work reported here focuses on the parsing and indexing of the audio stream.

The immediate goal of the audio processing is to identify transition points between segments and to do an initial content oriented labeling of the segments. We use a combination of signal processing techniques for feature extraction, and "intelligent" symbolic level processing for decision making. The two processes work together in the sense that some hypotheses are formed based on initial signal processing work which in turn controls further signal processing to test, verify or refine the initial hypotheses.

The symbolic processing includes knowledge about characteristics of some of the basic signal types that we expect to encounter. Currently our data testbed is news stories, and the signals can be grossly classified as "music," "speech," and "other," for the purpose of delineating meaningful segments. Speech is then further broken into segments at boundaries between different speakers.

23.2 INITIAL PROCESSING

The signal is passed first through a filter which measures the amplitude envelope. Labels are attached to the signal identifying regions where energy is at some threshold percentage below the average where some of the following stages do not need to do any work. A 256-point FFT is then performed.

23.3 MUSIC

The "music detector" is an extension of the work by Hawley (Hawley, 1993). No deep philosophical issues about what music is are being addressed here. The system computes peaks in the magnitude spectrum, then bases its decision on the average length of time that peaks exist in a narrow frequency region. We improved previous work by using an ERB (Equivalent Rectangular Bandwidth) scaling of the frequency region (Moore & Glasberg, 1983). Because this scaling is log-like above 500 Hz, it tends to be more robust than a linear representation because the sensitivity to peak movement is more uniform across frequency when the fundamental frequencies of speech or pitched musical instruments are nonstationary. Music detection is performed early because signals so labeled need no further analysis for our purposes. Sections that are not labeled as music are mined for more information.

23.4 PITCH

A spectrally based pitch detection algorithm (Cohen, Grossberg & Wyse, 1995) is employed which was designed to model aspects of human pitch perception. Also based on an ERB scaled energy representation of the signal, it employs an excitatory-center, inhibitory-surround mechanism that enhances peaks, and a weighted summation of regions around harmonics, to derive an activation strength function across pitch.   It is robust under conditions of mistuned components and models human responses to rippled noise and noise-band edges in addition to simple harmonic complexes.

The pitch model is layered, and includes a spectral representation, a contrast-enhanced spectral representation and finally a pitch layer, where, in general, every pitch has some level of activation. The pitch detector is robust against the effects of certain kinds of noise.   Broadband noise is ignored, for example, even when the signal to noise ratio (in dB) is negative. Due to the convolution with the "Mexican hat" on-center off-surround kernel, spectrally broad signals are suppressed before influencing the pitch layer. More compact signals, particularly those with energy across several harmonically related components, are represented in the pitch layer, but unless they are specially constructed to do so, tend not to shift the peaks due to other pitched signals. This robustness to noise does not make this a model of multiple source segregation, however. Even a single tone creates many peaks in the pitch layer (at all subharmonics, though only one is maximal), so there is no obvious way to associate source perception with any but the most salient peak.

In order to track the pitch of voiced speech over time, several auditory streaming constraints (Bregman, 1990) are embedded in a following processing stage. Because the pitch detector responds to peaked and rippled noise, noise from fricatives causes peaks to appear in the pitch activation function, and, especially if the fricative is unvoiced, a noise peak can be the most prominent one. The resulting trace of the maximally activated pitch makes jumps that are too far in frequency and too fast in time for humans to track as a single stream. By incorporating constraints concerning the relationship between the distance and the rate of frequency jumps that result in a sequence of tones either streaming together or breaking into several streams, we are able to keep the pitch tracker following the pitch of just the voiced portion of speech. Similar constraints concerning energy keep the tracker from being distracted by low level pitched sounds or brief nonspeech bursts.

23.5 SPEECH LABELING

At this point in the pipeline, we have several representations of the signal and a stream of time-stamped labels. To label a segment as "speech," the next stage examines the pitch track in segments not already carrying a label incompatible with speech. The speech label begins with a pitched (assumed now to be "voiced") segment. The label ends with the last pitched segment before a time interval greater than one second in which no pitched segment lasted more than 75 ms. These criteria were empirically determined.

23.6 SPEAKER DISCRIMINATION

Speaker discrimination is an important component of segmenting an audio stream into meaningful subunits. Understanding when speakers change is crucial for dialogue understanding. In the realm of newscasts, a change in speaker almost always corresponds to a change in the content of the news story. Speaker discrimination is related to speaker identification and verification, but the latter two processes are based on prior knowledge about a limited number of speaker identities and are usually text dependent. In speaker discrimination, only knowledge about speech in general is embedded in the system which is text independent. For the discrimination task, no matching of different segments to templates is done, only temporally local decisions about speaker changes are made. Despite the fact that "interspeaker variation" is the bane of speech recognition, actually extracting features that are invariant for one speaker, and that differ across speakers, is a challenging task.

Humans manage to recognize a change in speaker in a very short time, so averaging measures across tens of seconds should not be necessary. The methods used in our system combine pitch and spectral features and make use of timing cues as well. Before the discrimination processes run, a segment must first be labeled as speech. Potential speaker transitions are flagged by events such as lengthy segments of nonspeech, or sudden changes in pitch. Spectral features are extracted which are used for the final label assignment.

23.6.1 Pitch-Based Speaker Discrimination

Changes in pitch characteristics make an important contribution to speaker discrimination, but are neither necessary nor sufficient for identifying the transition. The cue is perhaps most reliable when the transition is between speakers of different gender, but the overlap of ranges is still considerable. Male vocal chords, tending to be longer and heavier than females', generally produce fundamental frequencies in the range between 80 Hz and 250 Hz, whereas those produced by females are generally in the range between 150 Hz and 500 Hz. The range for children is slightly higher than that for women.

Averaging of the speech signal over a window of time and looking for large changes in this measure is a possible technique, but we have found that pitch in a single utterance can vary widely even when averaged over a window of two or more seconds. Averaging also has the disadvantage of being too influenced by extremes; the more outlandish, the greater the influence. We have therefore adopted the use of a change in pitch range for flagging possible speaker transitions.

The range has two frequency bounds: one above which a certain percentage of input pitch values lie, the other below which a certain percentage lie over the duration of a time window. If a cutoff percentage parameter is set at 50%, for example, the mean is tracked. We are currently using a cutoff of 25% for both the upper and lower range, and a window of 2 seconds. The actual frequency of outliers thus have no effect on the range computation no matter how outlandish, making this method more robust than averaging with particularly "prosodic" speakers.

The temporal localization ability of the range change discrimination technique is better than the window size, since it depends upon the cutoff percentages as well. With cutoff percentages of less than 50%, the high bound is more sensitive (responds more quickly) to an increase in upper range than to a decrease, and the low bound is more sensitive to a decrease in the lower range than an increase. Changes to the range in the "sensitive direction" of the bound measures can happen as quickly as the cutoff percentage multiplied by the window length.

23.6.2 Spectrally Based Speaker Discrimination

Speaker variation is the bane of speech processors, and great pains are taken to normalize, compensate or otherwise make systems less sensitive to them. Sources of variation include regional accents, emotional stress, speaking rate, physical impediments, health, gender, and age, and chest, glottis, and vocal tract morphology. With so much interspeaker variability, it seems that automatic speaker discrimination should be easy. The difficulty, of course, lies in finding acoustic features that change less with one speaker than across many speakers.

We are currently exploring a spectrally based method. It is even more evident for spectra than for pitch that no average over a window short enough for reasonably fast detection of speaker change will be stationary over the course of a single speaker utterance.

One way to eliminate the effects of intraspeaker spectral variation would be to compare particular phonemes of one speaker to the same phonemes of another (recall that phonemes are linguistically, not acoustically, defined). This is one of the methods used in speaker identification and verification when comparing input to a known stored utterance (Furui, 1986). There are several problems with this approach. First, it involves the identification of the phonemes.

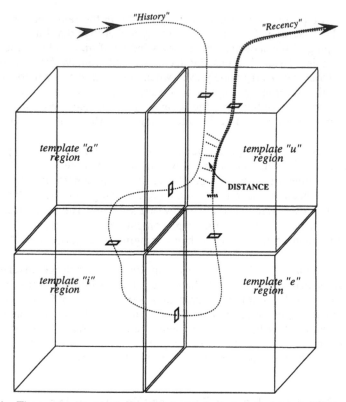

FIG. 23.1. The successive cepstral vectors trace a trajectory in the tessellated space, and vectors within the "recency" window (most recent .75 seconds) are compared to those in the "history" window (extending back 3 seconds) that were in the same template region. If the distance between a recency vector and the closest history vector exceeds a threshold, it adds to the novelty score.

Because one of the manifestations of speaker differences is that different vowels spoken by the respective speakers can overlap in formant space (Peterson & Barney, 1952), then a fairly complete speech recognition system would have to be a part of the discrimination mechanism and would thus carry a substantial computational burden. Another problem with this approach is that speaker discrimination would be language dependent, and people can normally detect speaker changes even when a language unknown to the observer is being spoken.

Our method, related to this same-phoneme comparison method, is to break up a spectral space into regions and compare new input only to stored data in the same spectral region that the input belongs to. This eliminates the need for phoneme identification (though at the expense of being able to use that particular aspect of variation in the process), and turns the approach into a kind of spectral redundancy measure.

To break the representation space into regions, we recorded 15 different voiced sounds (vowels, liquids and fricatives) at as close to a steady state as they could be spoken. Sixteen LPC-derived cepstral coefficients were taken using 25 ms windows stepped every 10 ms, and the vectors were averaged to produce one representative vector for each sound.

During the processing of segments of the input stream that have already been labeled by our system as both "speech" and "pitched," the most recent 750 ms (the "recency" window) of input vectors are compared to the previous 3 seconds (the "history" window), and a novelty score is computed. If the novelty score exceeds a certain threshold, then a new speaker is flagged as starting at the time corresponding to the beginning of the recency window.

The way the novelty score is computed is by first identifying the region into which each input vector falls, and then finding the closest vector in the history window already stored in that region. Euclidean distance, a standard for comparing cepstral vectors, is used. If this distance exceeds a threshold parameter, then the input vector is flagged as novel (see Figure 23.1). If the number of vectors in the recency window that are flagged as novel is greater than a second threshold parameter (expressed as a percentage of the number of data points that the recency window holds), then the criteria for identifying a speaker transition is met.

A brief description of how the parameters were determined sheds some light on how the method works. The tessellation of cepstral space limits the range of history vectors that recent inputs are compared to. This is what prevents the intraspeaker spectral variation, due to different vowels, to influence the measure. Thus the number of regions must be large enough to prevent too many cross-vowel comparisons. Our tessellation corresponds roughly to the number of vowels and semivowels (glides and liquids) used in the English language. The number of regions must not be too large, lest there never be a "history" vector in the same region as the input for comparison. Similarly, the length of the history window, although needing to be as short as possible to achieve acceptable temporal resolution, needs to be long enough so that at any given time, there is a high probability that there are history window vectors in the same region as the input. The 3 seconds of "history" maintained provides reasonable assurance that much of the input will fall in regions with stored vectors. The recency window needs to be short for resolution, but long for robustness, and long enough so that history vectors in the same region are from a previous entry of the trajectory into the region. When the recency window is so short that this condition is not met, then the distance between a recency vector and a history vector is determined by their distance along the trajectory itself rather than a speaker-characteristic use of the region of cepstral space.

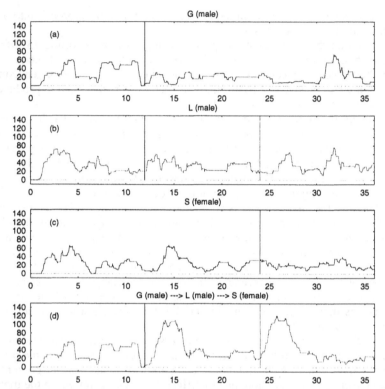

FIG. 23.2. Novelty scores for 36 seconds of a paragraph read by (a) male speaker "G," (b) male speaker "L," and (c) female speaker "S." (d) The novelty score for the same paragraph spliced together from the first 12 seconds from "L," the second 12 seconds from "G," and the last 12 seconds from "S." The traces of the individual speakers can be recognized in (d) except just after the speaker changes where the score suddenly jumps.

This spectral discrimination component of the system is still being developed, but preliminary results show some promise. Figure 23.2 (a,b,c) shows the running novelty score for 3 different speakers reading the same passage from a book. The maximum possible novelty score is 150 (if each of the 5 ms-spaced vectors in the .75 second recency windows was novel). The input for Figure 23.2 (d) was constructed from the first 12 seconds from the first reader, the second 12 seconds from the second reader and the final 12 seconds from the third reader. At each splice, the speaker changed in midsentence (though not midword), whereas the natural flow of the text was maintained. The novelty scores following each speaker changes can be seen to reach a peak higher than any of the individual speaker scores.

Preliminary investigations suggest that the method described is fairly robust, although the variations in a single speaker can still produce novelty scores in the vicinity of speaker-change peaks. The technique has the advantage of having a relatively light computationally load since it requires no speech or phoneme recognition, and because the input vector is only compared to a fraction of the recent speech utterance. The computation other than the distance measures is minimal. The method might also be useful for text-independent speaker identification and verification by concatenating stored speech of a known speaker with a new speech signal and deriving the novelty score.

There are a number of ways in which this spectral method might be improved. Using Perceptual Linear Predictive (PLP) analysis rather than Linear Predictive Coding (LPC) derived cepstral coefficients may prove beneficial since distances between PLP vectors have been shown to correlate more consistently with perceptual distance (Hermansky, 1990). The regions could be made adaptive with some continuously updated clustering method. Although most likely improving the performance over the a priori and arbitrary division of the representation space, this would add considerable computation time. It also appears that some of our representation regions are proving to be more useful for discrimination than others (which is suggested by the observation that parts of the spectrum are more useful than others in identifying voice types (Bloothooft & Plomp, 1986), and a systematic exploration of this will undoubtedly improve both the speed and the accuracy of this technique. Finally, other acoustic features that typically vary across speakers, but have a high intraspeaker variability with a high correlation to spectral variation (e.g., spectral tilt, relative levels of even and odd harmonics, breathiness) might be more usefully compared by region in the manner described herein.

23.7 DISCUSSION

In our video classification system, segment transition decisions in audio are based on less temporally localized information than are video transition decisions. However, all event labeling is done within 2 seconds of the event, and the processing runs in close to real time on a Sparc workstation, the actual run time being signal dependent.

The whole audio subsystem consists of some 20 or so different signal and symbol processing "filters" which can be run in a flexible ordering depending on the goals of the system. A uniform method of labeling and communication between processes has been developed which allows the information gleaned from one processes to control the processing parameters of another. Future work along these lines will make the flow of processing through the different filtering processes more flexible and integrated.

REFERENCES

Bloothooft, G. & Plomp, R. (1986). Spectral analysis of sung vowels, III. Characteristics of singers and modes of singing. *J. acoust. Soc. Am.*, 79(3), 852-864.

Bregman, A. (1990). *Auditory Scene Analysis*. Cambridge, MA: MIT Press.

Cohen, M., Grossberg, S., & Wyse, L. (1995). A spectral network model of pitch perception. *J. Acoust. Soc. Am.*, 98(2), 862-879.

Furui, S. (1986). Research on individuality features in speech waves and automatic speaker recognition techniques. *Speech Communication*, 5(2), 183-197.

Hawley, M. J. (1993). *Structure out of Sound*. Ph.D. Thesis, MIT

Hermansky, H. (1990). Perceptual linear predictive (PLP) analysis of speech. *J. Acoust. Soc. Am,.* 87(4), 1738-1752.

Moore, B., & Glasberg, B. (1983). Suggested formulae for calculating auditory filter bandwidths and excitation patterns. *J. Acoust. Soc. Am,.* 74, 750-753.

Peterson, G. E. & Barney, H. (1952). Control methods used in a study of vowels. *J. Acoust. Soc. Am,.* 24(2), 175-184.

Smoliar, S. W., & Zhang, H. J. (1994). Content-based video indexing and retrieval. *IEEE Multimedia*, 1(2), 62-72.

24

Using Musical Knowledge to Extract Expressive Performance Information from Audio Recordings

Eric D. Scheirer
The MIT Media Laboratory

A computer system is described which performs polyphonic transcription of known solo piano music by using high-level musical information to guide a signal-processing system. This process, which we term expressive performance extraction, maps a digital audio representation of a musical performance to a Musical Instrument Digital Interface (MIDI) representation of the same performance using the score of the music as a guide. Analysis of the accuracy of the system is presented, and its usefulness both as a tool for music-psychology researchers and as an example of a musical-knowledge-based signal-processing system is discussed.

24.1 INTRODUCTION

Traditionally, transcription systems (computer systems that can extract symbolic musical information from a digital-audio signal) have been built via signal processing from the bottom up. In this chapter, I examine a method for performing a restricted form of transcription by using a high-level music-understanding system to inform and constrain a signal-processing algorithm.

The motivation for this work was to build a system that could extract performance parameters from digital audio recordings of known solo piano music accurately enough that they could be used for music–psychological analysis of expressively performed music. However, the system also stands as a example of a new general approach to the polyphonic transcription problem.

The parameters extracted are those that are controllable by the pianist: velocity (the force of the stroke), and attack and release timing. Palmer (Palmer, 1989) suggested certain levels of timing accuracy in performance which can be understood as benchmarks for a system which is to extract note information accurately enough to allow the study of musical interpretation. For example, among expert pianists, the melody of a piece of music typically runs ahead of its accompaniment; for chords, where it is indicated that several notes are to be struck together, the melody note typically leads by anywhere from 10-15 ms to 50-75 ms, or even more, depending on the style of the music. Thus, if we are to be able to use such a system for examine timing relationships between melodies and harmony, it must be able to resolve timing differences between "simultaneous" notes to this level of accuracy or finer.

Five ms is generally taken as the threshold of perceptual difference (JND) for musical performance (Handel, 1989); if we wish to be able to reconstruct identical performances from the transcription data, the extracted timing data must be at least this accurate.

Score-Based Transcription; or, Why Cheating Is Good. It seems on the surface that using the score to aid transcription is "cheating," or worse, useless – what good is it to build a system that extracts information you already know? It is our contention that this is not the case; and, in fact, score-based transcription is an extremely useful restriction of the general transcription problem.

It is clear that the human music-cognition system is working with representations of music on many different levels which guide and shape the perception of a particular musical performance. Work such as Krumhansl's tonal hierarchy (Krumhansl, 1991) and Narmour's multilayered grouping rules (Narmour, 1990), (Narmour, 1993) show evidence for certain low- and midlevel cognitive representations for musical structure; and syntactic work such as Lerdahl and Jackendoff's (Lerdahl & Jackendoff, 1983), while not as well-grounded experimentally, suggests a possible structure for higher levels of music cognition.

Although the system described in this chapter does not attempt to model the human music cognition system per se (and, further, it is not at all clear how much transcription the human listener does, in the traditional sense of the word – see section 24.4.2, below), it seems to make sense to work towards multilayered systems that treat musical information on a number of levels simultaneously. This idea is similar to those presented in Oppenheim and Nawab (Oppenheim & Nawab, 1992) regarding symbolic signal processing.

From this viewpoint, score-aided transcription can be seen as a step in the direction of building musical systems with layers of significance other than simply a signal-processing network. Systems built along the same lines with less restriction might be rule-based rather than score-based, or even attempt to model certain aspects of human music cognition. Such systems would then be able to deal with unknown as well as known music.

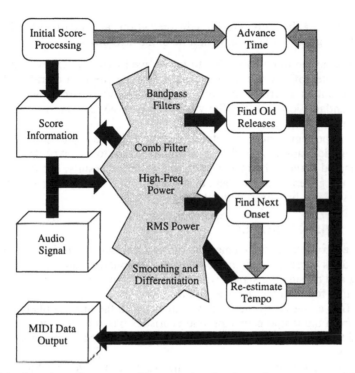

FIG. 24.1. Overview of system architecture, showing the main components (algorithms and data sources) of the system and how they interact.

24.2 ARCHITECTURE OF THE SYSTEM

Figure 24.1 shows a schematic representation of the architecture of the current implementation of the computer program.[1]

Briefly, the structure is as follows: A initial score- processing pass determines predicted structural aspects of the music, such as which notes are struck in unison, which notes overlap, and so forth. In the main loop of the system, we do the following:

• Find releases and amplitudes for previously discovered onsets.

• Find the onset of the next pitch in the score.

- Re-examine the score, making new predictions about current local tempo in order to guess at the location in time of the next onset.

Once there are no more onsets left to locate, we locate the releases and measure the amplitudes of any unfinished notes and write the data extracted from the audio file out as a MIDI text file. It can be converted using standard utilities into a Standard Format MIDI file which can then be resynthesized using MIDI hardware or software.

It is important to note the relative simplicity of the signal- processing in the following descriptions of the algorithms. It is to be expected that more sophisticated filter techniques would lead to better results, but it is a mark in favor of the general attractiveness of the use of high-level information that good performance can be gained without them.

24.2.1 Onset Extraction

The onset extractor is the element of the system into which the most work has gone and, as a result, is the most complex. It contains four main methods for filtering and parsing the signal, with parameters that are adjustable based on the information found in the signal and in the score-processing.

To find an onset, a time window within which the onset is likely to occur is passed from the tempo-estimator, and heuristics based on the score information are used to determine the type of digital signal processing which should be done, as follows.

If no other notes are struck at the same time, we can look for high-frequency energy (above 4000 Hz; this energy is noise from the hammer strike) in the signal during this time window, and an increase in the overall RMS power. If either of these occur in an area of positive derivative in the fundamental band of the pitch we are looking for, we can select the point of peak derivative of the high-frequency energy or RMS power as the onset time. This method leads to the highest accuracy (see Results in the Validation Experiment section for quantitative analysis of the accuracy).

If we cannot find a suitable peak in the high-frequency or RMS power, we use a comb filter based on the fundamental frequency of the target pitch. We look for the sharpest peak in the derivative of the RMS of the filtered signal, which will generally correspond to a point in the middle of the rise to peak power. We slide back in time to find the positive-going zero-crossing in the derivative, and take this as the onset. This introduces a timing bias into the extraction, but it can be corrected through comparison to a "ground truth" signal to build a statistical model of the average bias by pitch.

In the case where multiple notes are struck simultaneously, we cannot use high-frequency or RMS power to select onset points, because there is no way to tell if the energy burst corresponds to a particular note in a chord, or to the chord as a whole. Instead, we build a multiple-bandpass filter, with pass regions selected to be those harmonics of the current pitch that do not have interference from another pitch – that is, that are not also overtones or fundamentals of another note that occurs at the same time. We filter the signal using this multiple-bandpass and take the RMS, and then use the derivative-based estimator described above.

The Q (ratio of center frequency to bandwidth) of the filters is selected depending on the expected proximity of other notes in pitch and time, ranging from 15 to 50. There is bias implicit in the higher group delay of narrowly tuned filters; this is a possible source of error in the algorithms which should be examined more closely.

24.2.2 Release Timing and Amplitude Measurement

Release timing and amplitude measurement is done with simpler techniques than the onset extraction. We build a multiple- bandpass filter based on the harmonics we know to be usable from the score-information, in a time window going from the previously extracted onset time to a point 3 sec later in the signal. We look forward in time for the peak-power point in the filtered signal, and extract that as the amplitude. We then continue to look forward, for the point at which the power either drops below 10% of the peak power, or begins rising to another peak. This point is extracted as the onset. (See the Discussion section for criticisms of this method.)

24.2.3 Tempo Estimation

Tempo re-estimation is performed each time through the main program loop, to attempt to understand the local timing of the current performance and derive good guesses for the locations of the next few note onsets.

Currently, a regression line is calculated, matching predicted onset time for the last ten notes from the score against their extracted onset times. Predictions are then made by using the regression line to extrapolate timings for the next five notes. When the first note of a chord has been extracted, however, we choose its time as the prediction for the other notes in that chord.

This method is adequate for following the performance of the pieces used in the validation experiment. There are, of course, many other techniques for robust performance-following in the literature – (Vercoe, 1984), for example.

24.3 VALIDATION EXPERIMENT

To analyze the accuracy of the timing and velocity information extracted by the system, a validation experiment was conducted using a Yamaha Disclavier MIDI-recording piano. This device has both a conventional upright piano mechanism, enabling it to be played as a standard acoustic piano, and a set of sensors that enable it to capture the timings (note on/off and pedal on/off) and velocities of the performance in MIDI format. The Disclavier also has solenoids

FIG. 24.2. Musical examples used: (top) Fugue 16, in G minor, from Book I of the Well-Tempered Clavier by J.S. Bach; (bottom) the first piece "Von fremden ländern und menschen" from Schumann's Kinderszenen Suite.

that enable it to be used to play back prerecorded MIDI data like a player piano, but this capability was not used.

Scales and two excerpts of selections from the piano repertoire were performed on this instrument by an expert pianist; the performances were recorded in MIDI using the commercial sequencer Studio Vision by Opcode Software, and in audio using Schoeps microphones. The DAT recording of the audio was copied onto computer disk as a digital audio file; the timing-extraction system was used to extract the data from the digital audio stream, producing an analysis that was compared to the MIDI recording captured by the Disclavier.

It is assumed for the purposes of this experiment that the Disclavier measurements of timing are perfectly accurate; indeed, it is unclear what method could be used to evaluate this assumption in a robust fashion. One obvious test, that of resynthesizing the MIDI recordings into audio, was conducted to confirm that the timings do not vary perceptually from the note timings in the audio, and this was in fact found to be the case.

24.3.1 Performances

There were 8 musical performances, totaling 1005 notes in all, that were used for the validation experiment. Three were scales: a chromatic scale, played in quarter notes at m.m. 120 (120 quarter notes per minute) going from the lowest note of the piano (A four octaves below middle C, approximately 30 Hz) to the highest (C three octaves above middle C, approximately 4000 Hz); a two-octave E-major scale played in quarter notes at m.m. 120; and a four-octave E-major scale played in eighth notes at m.m. 120. Each of the two E-major scales moved from the lowest note to the highest and back again 3 times.

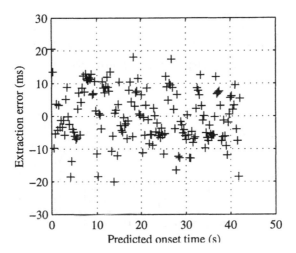

FIG. 24.3. Predicted vs. extracted onset times. In general, onset extraction error is small.

Additionally, three performances of excerpts of each of two pieces, the G-minor fugue from Book I of Bach's Well-Tempered Clavier, and the first piece "Von fremden Ländern und Menschen" from Schumann's Kinderszenen Suite, op. 15, were recorded. The first line of the score for each of these examples is shown in Fig. 24.2. All three Bach performances were used in the data analysis; one of the Kinderszenen performances was judged by the participating pianist to be an poor performance, suffering from wrong notes and unmusical phrasing, and was therefore not considered.

These pieces were selected as examples to allow analysis of two rather different styles of piano performance: The Bach is a linearly constructed work with overlapping, primarily horizontal lines, and the Schumann is vertically oriented, with long notes and heavy use of the damper pedal.

24.3.2 Results

Figures 24.3 to 24.11 show selected results from the timing experiment. We deal with each of the extracted parameters in turn: onset timings, release timings, and velocity measurements. In summary, the onset timing extraction is successful, and the release timing and amplitude measurement less so. However, statistical bounds on the bias and variance of each parameter can be computed which allow us to work with the measurement to perform performance analysis of a musical signal in the stochastic sense.

Onset Timings. Foremost, we can see that the results for the onset timings are generally accurate. Figure 24.3 shows a scatterplot of the predicted onset time (onset time as recorded in the MIDI performance) versus extraction error (difference between predicted and extracted onset time) from one of the Schumann performances. The results for the other pieces are similar.

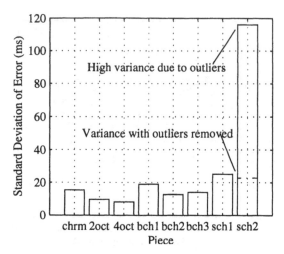

FIG. 24.4. Onset error standard deviation for each performance. Except for one piece with "outlier" errors, there is little difference between system performance for differing musical styles.

This is not nearly a strict enough test for our purposes, however. One possibility is to resynthesize the extracted performances and compare them qualitatively to the originals; or, for a quantitative comparison, we can examine the variances of the extracted timing deviations from the original.

Treating a piece as a whole, there is not useful information present in the mean of the onset timing deviations, as this largely depends on the differences in the start of the "clock time" for the audio vs. MIDI recordings; measuring from the first onset in the extraction and the first attack in the MIDI simply biases the rest of the deviations by the error in the first extraction. In fact, the first extraction is often less accurate than those part way through the performance, because there is not a tempo model built yet.

Thus, the global data shown below deals only with the variance of extraction error around the mean extraction "error". However, for results dealing with subsets of the data (i.e., only monophonic pitches, or only pitches above a certain frequency), there are useful things to examine in the mean extraction error for the subset relative to the overall mean extraction error. We term this between-class difference in error the bias of the class

Fig. 24.4 shows the standard deviation of onset timing extraction error for each of the eight pieces used (in order, the chromatic scale, the two-octave E major scale, the four-octave E major scale, the three performances of the Bach, and the two performances of the Schumann). We can see that the standard deviation varies from about 10 ms to about 30 ms with the complexity of the piece. Note that the second performance of the Schumann excerpt has an exceptionally high variance. This is because the tempo subsystem mispredicted the final (rather extreme) ritardando in the performance, and as a result, the last

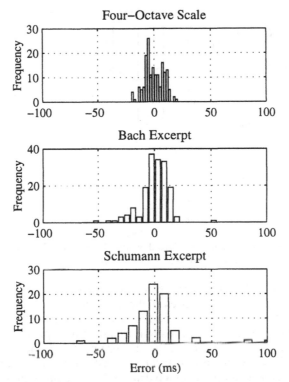

FIG. 24.5.   Onset error standard deviation for three performances. The errors are distributed roughly normally, which implies that a Gaussian noise model is appropriate for stochastic characterization of the output of the system.

five notes were found in drastically incorrect places. If we throw out these outliers as shown, the variance for this performance improves from 116 ms to 22 ms.

Fig. 24.5 shows histograms of the deviation from mean extraction error for a scale, a Bach performance, and a Schumann performance. For each case, we can see that the distribution of deviations is roughly Gaussian or "normal" in shape. This is an important feature, because if we can make assumptions of normality, we can easily build stochastic estimators and immediately know their characteristics. See the Discussion section for more on this topic.

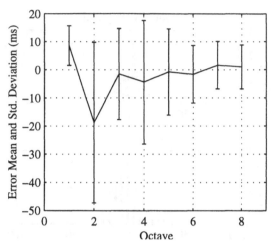

FIG. 24.6. Onset error mean and standard deviation by octave. Note bias of error by pitch, and increased accuracy (lower variance) for higher pitches.

We can also collect data across pieces and group it together in other ways to examine possible systematic biases in the algorithms used. Figure 24.6 shows the bias (mean) and standard deviation of onset timing extraction error collected by octave. We see that there is a slight trend for high pitches to be extracted later, relative to the correct timing, than lower pitches. Understanding this bias is important if we wish to construct stochastic estimators for the original performance. Note that this is not a balanced data set; the point in the center-of-piano octave represents about 100 times more data than the points in the extreme registers.

Similarly, Figure 24.7 shows the bias and standard deviation of onset timing extraction error collected by the method used to extract the onset. As discussed in the Algorithms section, different methods are used to extract different pitches, depending upon the characteristics of the high-level score information, and upon the heuristic information extracted by the signal processing networks.

In Fig. 24.7, the "method used" is as follows:

- No notes were struck in unison with the extracted note, and there is sufficient high-frequency energy corresponding with positive derivative in the fundamental bin to locate the note.

- No notes were struck in unison with the extracted note. High frequency energy could not be used to locate the note, but RMS power evidence was used.

- No notes were struck in unison with the extracted note; but there was not sufficient high frequency or RMS evidence to locate the note. The comb-

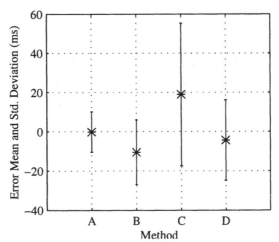

FIG. 24.7. Onset error mean and standard deviation by extraction method. See text for key to methods. As expected, method A (high-pass energy) is the most accurate and least biased.

filter and derivative method was used. These are, in general, represent "hard cases," where the audio signal is very complex.

- There were notes struck in unison with the extracted note, so high-frequency and RMS power methods could not be used. The allowable overtones and derivative method were used.

We can see that there is a bias introduced by using method C, and relatively little by other methods. In addition, it is clear that the use of the high-frequency energy or RMS power heuristics, when possible, leads to significantly lower variance than the filtering-differentiation methods.

Release Timings. The scatterplot of predicted release timing is shown in Figure 24.8. As can be seen, there is similarly high correlation between predicted and extracted values as in the onset data. We can also observe a time relation in the data – this is due to the bias of release timing by pitch.

We can additionally plot predicted duration vs. extracted; we see that there is not nearly as much obvious correlation, although the r = .3163 value is still highly significant statistically. This is shown in Figure 24.9.

Amplitude/Velocity. A scatterplot of predicted velocity against extracted log amplitude relative to the maximum extracted amplitude is shown in Figure 24.10. As with duration, there is a high degree of correlation in the data, with r = .3821, although obviously not as much as with the onset timing extraction.

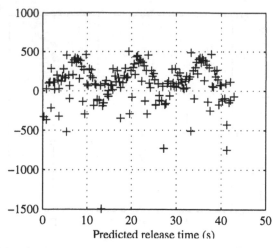

FIG. 24.8. Predicted vs. extracted release time. Bias over time corresponds to bias by pitch.

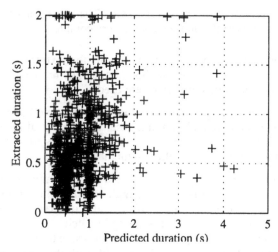

FIG. 24.9. Predicted vs. extracted duration. The data fits a linear model rather well (see text).

We can correct for the unit conversion between abstract MIDI "velocity" units in the predicted data and extracted log amplitude energy values by calculating the regression line of best fit to the fig 24.10 scatter-plot – $y = 7.89 - 79.4x$ – and using it to rescale the extracted values.

When we treat the amplitude data in this manner, we see that once again, the noise from extraction error is quite well represented as a Gaussian

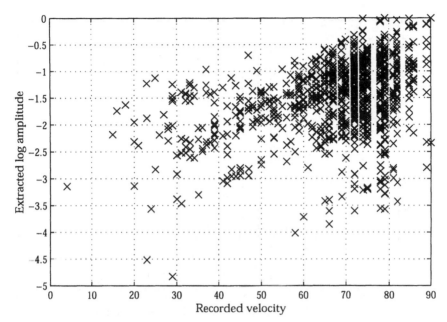

FIG. 24.10. Predicted MIDI velocity versus extracted amplitude. This data can be used to calculate the proper scaling of output velocities.

distribution (Figure 24.11), with standard deviation of error equal to 13 units on the MIDI velocity scale.

24.4 DISCUSSION

There are a number of different levels on which this work should be evaluated: as a tool for music-psychology research, as an example of a system that performs musical transcription, and as an example of a multilayered system that attempts to integrate evidence from a number of different information sources to understand a sound signal. We consider each of these in turn, and then discuss ways in which the current system could be improved.

24.4.1 Stochastic Analysis of Music Performance

Part of the value of the sort of variance-of-error study conducted in the Results section is that we can treat extracted data as a stochastic estimator (cf. for example, (Papoulis, 1991)) for the actual performance and make firm enough assumptions about the distribution of the estimation that we can obtain usable results.

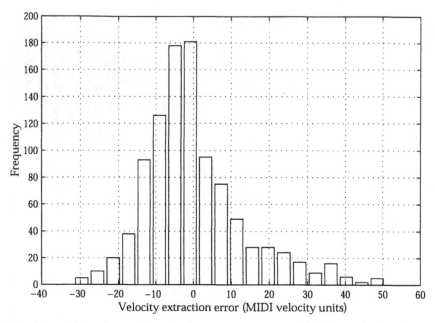

FIG. 24.11. Histogram of rescaled velocity extraction error. Again, note generally normal distribution.

It is clear that some aspects of expressive music performance can be readily analyzed in the constraints of the variance in extraction discussed above. For example, tempo is largely carried by onset information, and varies only slowly, and only over relatively long time-scales, on the order of seconds. Even the worst-case performance, with standard deviation of extraction error about 30 ms, is quite sufficient to get a good estimate of "instantaneous tempo" at various points during a performance.

For example, assume that two quarter notes are extracted with onsets 1.2 seconds apart, say at $t_1 = 0$ and $t_2 = 1.2$ for the sake of argument. We can assume, then, that these extractions are taken from Gaussian probability distribution functions (pdf's) with standard deviations of .02 seconds, and calculate the pdf of the inter-onset time $t_2 - t_1$ as Gaussian with mean 1.2 seconds and standard deviation .0283 seconds, giving us 95% probability that the actual tempo is in the interval [47.75, 52.48].

We can similarly recreate other sorts of analyses such as those found in (Palmer, 1989) or (Bilmes, 1993) by treating the timing variables as random Gaussian variables rather than known values.[2]

Depending on which question we want to answer, though, the answers may be less satisfactory for small timing details. For example, an important characteristic of expressive performance of polyphonic music is the way in which a melody part "runs ahead" or "lags behind" the accompaniment. To examine this question, we wish to determine the posterior probability that a particular note in a chord has been struck last, given the extracted onset timings.

Consider a two-note dyad, where the score indicates the notes are to be struck simultaneously; the onsets have been extracted as 1.000 and 1.015 sec, respectively. We can calculate the probabilities that the notes were actually struck within the 5 msec window of perceptual simultaneity, or that the earlier or later was, in fact, struck first. To do this calculation, we build a Bayesian estimator of the time lag, and use error functions; we find that the probability that the earlier extraction was actually struck first is 0.6643, that and that the later extraction was actually first is .2858, assuming that the standard deviation is the worst-case of 25 msec.

24.4.2 Polyphonic Transcription

It is clear that using this sort of layered method with the score enables polyphonic transcription with more accuracy than previously-existing systems. When the extracted MIDI is resynthesized, the resulting performance is clearly the same piece performed in the "same style;" it is not indistinguishable from the original performance, due to errors, but many of the important aspects of the original performance are certainly captured.

The system has not been exhaustively tested on a wide variety of musical styles. The Bach example has four-voice polyphony in the score, which ends up being six- or eight-voice polyphony at points due to overlap in the performance. The Schumann has heavy use of the damper pedal, and so has sections where as many as 9 notes are sustaining at once. The only musical cases that are not represented among the example performances analyzed above are very dense two-handed chords, with 6 or 8 notes struck at once, very rapid playing, and extreme use of rubato in impressionistic performance.

It is anticipated that any of these situations could be dealt with in the current architecture, although the tempo-follower would have to be made more robust in order to handle performance which are not well-modeled by linear tempo segments. This is generally a solvable problem, however – see (Vercoe, 1984) for an example.

A larger issue regarding the problem of general polyphonic transcription is the goal and motivations underlying them. Why is there so much interest in building transcription systems?

We submit that it is for several reasons. Obviously, having a working transcription system would be a valuable tool to musicians of all sorts – from music psychologists to composers (who could use such a tool to produce "scores" for analysis of works of which they had only recordings) to architects of computer music systems (who could use it as the front-end to a more extensive music-intelligence or interactive music system).

Another reason that so much effort has been invested in the construction of transcription systems is that on the surface, it seems as though it "should be" possible to build them, because the necessary information "must be" present in the acoustic signal. Although this feeling seems to underlie much of the work in this area, it is so far drastically unjustified.

This point relates to a final reason, which is based on a hypothesis of the human music cognition system – that human listeners are doing something rather like transcription internally as part of the listening process. Stated another way, there is an implicit assumption that the musical score is a good approximation to the midlevel representation for cognitive processing of music in the brain.

It is not at all clear at this point that this hypothesis is, in fact, correct. It may well be the case that in certain contexts (for example, densely orchestrated harmonic structures), only a schematic representation is maintained by the listener, and the individual notes are not perceived at all – see (Scheirer, 1996). Because it is exactly this case that existing transcription systems have the most difficulty with, perhaps we should consider building transcription systems with other goals in mind than recreating the original score of the music.

In particular, building systems which can build human like representations from an audio stream would be adequate to supplant transcription for at least two of the three foregoing goals. For music psychologists, it is obviously only necessary to be able to extract parameters from music at a level similar to that possible by human performers; and for architects of interactive music systems, although it might be desirable to extract more detailed information, it is clearly (because humans are quite good at music performance with the ears they have) not necessary.

Even for composers, or jazz musicians who wish to produce fixed records of performances to study on paper, a representation which contains only the information a human perceives in the music (say, a melodic improvisation, a bass line, and a harmonic/rhythmic structure) would be quite an asset, even if it did not tell the complete story.

24.4.3 Evidence-Integration Systems

The evidence integration aspects of the system are the most novel, and at the same time, the least satisfying. It is very difficult to build architectures which allow the use of data from many sources simultaneously; the one for this system is perhaps not as sophisticated as it could be. For example, the current system does not have the ability to use knowledge discovered in the attack (other than the timing) to help extract the release. Similarly, it would be quite useful to be able to examine the locations of competing onsets and decays in the extraction of parameters for a note with overlapping notes.

At the same time, though, the success of the system in its current state is promising with regard to the construction of future systems with more complex architectures.

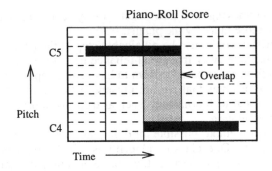

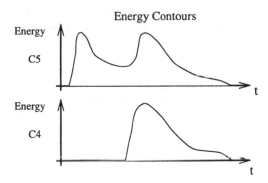

FIG. 24.12. A release gets buried by overlapping energy from a note an octave below, due to harmonic energy interference.

24.4.4 Future Improvements to System

There are many directions which contain ample room for improving the system. Obviously, more work is needed on the release- and amplitude–detecting algorithms. It is expected that more accurate amplitude information could be extracted with relatively little difficulty; the results here should be considered preliminary only, as little effort has currently gone into extracting amplitudes.

Release timings are another matter; they seem to be the case where the most sophisticated processing is required in a system of this sort. Figure 24.12 shows the major difficulty. When a note (for example, the C4 in Figure 24.12) is struck after but overlapping a note which has the fundamental corresponding to an overtone (the C5), the release of the upper note becomes "buried" in the onset of the lower. It does not seem that the current methods for extracting release timings are capable of dealing with this problem, and that instead, some method based on timbre-modeling would have to be used.

It would improve the robustness of the system greatly to have a measure of whether the peak extracted from the signal for a particular note has a "reasonable" shape for a note peak. Such a measure would allow more careful search and tempo-tracking, and also enable the system to recover from errors, both its own and those made by the pianist.

Such a heuristic would also be a valuable step in the process of weaning a system such as this one away from total reliance upon the score. It is desirable, obviously, even for a score-based system to have some capability of looking for and making sense of notes that are not present in the score. At the least, this would allow us to deal with ornaments such as trills and mordents, which do not have a fixed representation.

There are other methods possible for doing the signal- processing than those actually being used. One class of algorithms which might be significantly useful, particularly with regard to the above mentioned "goodness of fit" measure, is those algorithms which attempt to classify shapes of signals or filtered signals, rather than only examining the signal at a single point in time. For example, we might record training data on a piano, and use an eigenspace method to attempt to cluster together portions of the bandpass-filtered signal corresponding to attacks and releases.

Ultimately, it remains an open question whether a system such as this one can be expanded into a full-fledged transcription system which can deal with unknown music. Certainly, the "artificial intelligence" component, for understanding and making predictions about the musical signal, would be enormously complex in such a system. Work is currently in progress on a "blackboard system" architecture (see, e.g. (Oppenheim & Nawab, 1992)) for investigation of these issues. An initial system being built using this architecture will attempt to transcribe "unknown but restricted" music – the set of four-part Bach chorales will be used – by development of a sophisticated rule-based system to sit on top of the signal processing.

24.5 CONCLUSION

Results here show that certain limited aspects of polyphonic transcription can be accomplished through the method of "guess and confirm" given enough a priori knowledge about the contents of a musical signal.

The resulting system is accurate enough to be useful as a tool for investigating some, but not all, aspects of expressive musical performance. The uncertainty introduced into note timings as part of the extraction is small enough to allow accurate tempo estimation and perhaps certain sorts of studies of phrasing. The system in its current form is probably not yet accurate enough to investigate subtle questions of melodic–harmonic timing.

Possible future work includes:

(1) Increasing the accuracy of the algorithms used in this system.

(2) Exploring the use of other algorithms or timbre models to augment the algorithms already in place.

(3) Building heuristics to determine whether a candidate note in a signal is likely to actually be a note.

(4) Building systems that can extract from instruments other than pianos, with more complex envelope shapes.

(5) Building rule-based or music-cognitive systems to replace the role of the score in this system.

ACKNOWLEDGMENTS

Thanks to Barry Vercoe, Michael Hawley, and John Stautner for their advice during the progress of this research. Thanks also to Charles Tang for the piano performances used in the validation experiment, and to Teresa Marrin for providing the initial suggestion that led to this research. As always, the graduate students in the Machine Listening Group of the Media Lab have been helpful, insightful, provocative, and ultimately essential in support of the production of this paper and the research on which it is based.

REFERENCES

Bilmes, J. (1993). Timing is of the essence: Perceptual and computational techniques for representing, learning, and reproducing expressive timing in percussive rhythm. Master's thesis, MIT Media Laboratory.

Handel, S. (1989). *Listening*. Cambridge, MA: MIT Press.

Krumhansl, C. (1991). *Cognitive Foundations of Musical Pitch*. Oxford: Oxford University Press.

Lerdahl, F. & Jackendoff, R. (1983). *A Generative Theory of Tonal Music*. Cambridge, MA: MIT Press.

Narmour, E. (1990). *The Analysis and Cognition of Basic Melodic Structures*. Chicago: University of Chicago Press.

Narmour, E. (1993). *The Analysis and Cognition of Melodic Complexity*. Chicago: University of Chicago Press.

Oppenheim, A. & Nawab, S. H. (1992). *Symbolic and Knowledge-Based Signal Processing*. Prentice-Hall, Inc.

Palmer, C. (1989). *Timing in Skilled Music Performance*. Ph.D. thesis, Cornell University.

Papoulis, A. (1991). *Probability, Random Variables, and Stochastic Processes* (third edition). New York: McGraw-Hill.

Scheirer, E. (1996). Bregman's chimerae: Music perception as auditory scene analysis. In *Proc. 1996 Intl Conf on Music Perception and Cognition*.

Vercoe, B. (1984). The synthetic performer in the context of live performance. In *Proc. 1984 Int. Computer Music Conf.*

NOTES

[1] All program code is currently in the Matlab matrix-processing language, and is available from the author; email to eds@media.mit.edu with requests.

[2] It is arguable that they should have been treated this way in the cited work to begin with, because there is bound to be sensor noise coming into play.

Author Index

Subject Index

A

ABBOT speech recognition system, simultaneous-speaker sound corpus, 328, 329

Absolute refractory periods (PRA), cochlear nucleus level and amplitude modulation, 60–62

Abstract architecture, IPUS blackboard architecture, 106–107

Abstract salience, midlevel representation, 259

ACF, *see* Autocorrelation functions

Acoustic issues, beat tracking of real-time audio signals, 159–160

Acoustic objects, environmental speech recognition, 179, 180, 181

Acoustic patterns, auditory scene analysis, 1

Acoustic properties, environmental speech recognition, 179, 180, 181, 187–188, 190–191

Acoustons, environmental speech recognition, 178, 182–183, 184, 189

Active phase, segmentation network in auditory stream segregation, 73, 77

Adaptive perceptual systems, discrepancy directed diagnosis and discrepancies, 223–225

learning algorithm and examples, 225–228

parameterized signal processing algorithms, 219–223

related work, 216–217

sound understanding testbed, 217–219

status, 228–229

Agency, residue-driven architecture of sound stream segregation, 196, 198–199

Agent-based systems, developments, xi

Agent-pair, beat tracking of real-time audio signals, 162, 164, 168–171

AI, *see* Artificial intelligence

AM, *see* Amplitude modulation

Ambiguity

beat tracking of real-time audio signals, 159

perceptual in neural oscillator model, 95

Amplitude, expressive performance extraction, 365

validation, 371–373, 374

Amplitude modulation (AM)

estimation in auditory scene analysis, 238–239

identification by neural cells, 14

midlevel representation, 263, 264

processing

basic modeling, 60–63

implication for modeling, 66–68

psychoacoustic experiments, 63–66

Amplitude modulation transfer function (AMTF), cochlear nucleus level of basic modeling, 61

AMTF, *see* Amplitude modulation transfer function

Analysis unit, cocktail party processing, 247

387